MUD, MANHOLES, AND MACHETES

TRUE STORIES FROM THE LIFE OF A SURVEYOR ENGINEER

A HUMOROUS MEMOIR BY **RITCHEY MARBURY**

Copyrighted Material

Mud, Manholes, and Machetes: True Stories
from the Life of a Surveyor Engineer
Copyright © 2019 by Ritchey M. Marbury, III. All Rights Reserved.

ISBNs: 978-1-7331478-0-4 (Hardcover)
978-1-7331478-1-1 (Softcover)
978-1-7331478-2-8 (eBook)

Printed in the United States of America

This book is dedicated to all land surveyors, engineers, and those associated with them, as they face the challenges of mud, dirty sewers, swamps, snakes, alligators, heat, cold, computers, and endless paperwork, in order to make this world a better place, while working in a profession they love.

CONTENTS

Introduction ix

Chapter 1: Elementary School Days 1
    Start of SAMSOG 1
    Surveying as a Young Lad 3
    Summer Jobs 7
    Watermelons in the Early Morning 10
    House in the Middle of the Street 11
    Valdosta Detention Pond 12
    Machine-Gun Transit 14
    Let the Water Flow Through 16
    A Helping Hand 17
    Working with My Dad as a Child 18

Chapter 2: High School Days 21
    Lovic Carl Marbury 21
    Romantic Benefits of Surveying 23
    Fonda and the Sanitary Sewer 25
    Three Meetings 26
    Don't Mess with a Dead Alligator 27
    Hillsman Park 29
    Spiral Streets 31

Street Names 32
Sure Cure for Asthma 34
Water on Linen 37
Surveying Rivers and Creeks 39
Surveyor Spanks Contractor 42
The Surveyor and the Skull 43
Dad Helps Sick Competitor. 44
Interesting Characters 45

Chapter 3: College and Military 49
Albany Shopping Center—Midtown Mall 49
Darton State College 51
Lake Park Subdivision 54
Summers and Graduate School. 57
Sand Dunes. 62
My Son, No My Son 64
John Herndon Lost His Glasses 65
Paulin Altimeter. 66
Military 68

Chapter 4: Dad and I Are Partners 83
Home from the Military. 83
Master of City Planning 84
If You Have No Hill, Make a Hill. 89
Surveying in a Cow Pasture. 91
Albany Mall 92
Double Your Bill. 98
Working for a Crook 100
Water Tanks 102
Dad and the Flying Instructor 105
Landing Backwards. 106
Night Takeoff and Lost Power 108

CONTENTS

    Collections Are a Stinky Business 110
    Not-So-Original Marker. 113
    Miller Apartments. 115
    Snow Skiing in Vermont. 117
    Always Bless Your Food 119
    George Melton and Prayer. 120
    One-Legged, One-Eyed Dog. 121
    Nancy Cartmell, Karate Bookkeeper 123
    Our Best Collection Agency—the IRS. 124
    The Zoo at Chehaw. 126
    Rick and the Swimming Pool 128
    President Jimmy Carter 129
    Fun Working with My Dad 132
    Idaho Mission 137

Chapter 5: Survival After Dad's Retirement 139
    Starting Over with No Money. 139
    Archwood Drive 141
    Pecan Street—Cordele, Georgia 143
    Peanut Hulls in the Ditch. 145
    Muddy Motivation 147
    Watch That Machete—Don't Scalp My Son 148
    Certify Plat Not Correct. 149
    Pit Toilets in the Neighborhood 151
    Purchased Stevenson and Palmer. 153

Chapter 6: Marbury Engineering Begins Again. 155
    Up from the Ashes. 155
    Hidden Lakes 156
    Lancaster Village 160
    Callaway Lakes. 163
    Hurricane Andrew 164

Grand Island 166
Itchauway Plantation 169
Red and Green are Sometimes Black and White 171
Snakes and Stephen Hart 173
I Lost My Dad 175
Flood of 1994 176
Hose Bib and Water Faucet 181
Doug Wingate Recognizes Service 182
Rod Hutchinson and the Rowdy Client 183
Quality Seminars 184
Lowe's Fire of 1996 186
Joint Ventures 188
Puddle Expert 191
Inlets with No Outfall 192

Chapter 7: Life After Marbury Engineering Company 195
EMC Engineering 195
SRJ Engineering 200
Ritchey Marbury PE, RLS 202
City of Cordele, Georgia 204
First Cordele, Georgia, Assignments 206
Georgia Museum of Surveying and Mapping 210
Cordele, Georgia, Water Meter Survey 211
Cordele, Georgia, Hurricanes, and Tornados ... 213
In My Eighties, Working Full Time, and Loving It .. 217

About the Author 223

INTRODUCTION

This book is about how I motivated my survey crew by falling face first into a muddy swamp.

It is about how I, as a young surveyor, climbed out of a sanitary-sewer manhole, looked up, and saw my girlfriend, now my wife of more than 57 years, sitting on her front steps.

It is about how my son smiled after being cut on the head with a machete. I almost fainted and was taken from his hospital room in a wheelchair.

It is about how my wife obtained a degree from Georgia Tech without ever attending.

It is about how I landed my airplane in Tullahoma, Tennessee, going backwards.

It is about how I clung to the outside ladder of a water tank, one hundred fifty feet high, while strong winds blew me back and forth around the tank.

It is about how a rich woman raced her Cadillac toward my dad as he surveyed property near Radium Springs in Albany, Georgia.

It is about how a grouchy old man with a shotgun watched my dad help an old lady up the stairs.

It is about how one client told my dad to tear up his invoice and re-write it at double the amount, and he would pay it.

It is about how a contractor told me he never wanted to work on another of my projects again but wanted me to do all his engineering and surveying work.

It is about how, when Benny Harpe and I measured water-meter locations in Cordele, Georgia, I was attacked by five dogs.

It covers adventures with snakes, alligators, and other critters.

This is not a business book. It is not a textbook. It is not a technical book. This is a fun book of surveying and engineering stories. They are all true, and all are from my personal experiences as a land surveyor and professional engineer. It covers fun experiences from the time I was nine years old, through today, when I am in my eighties. It covers experiences working with my grandfather, my father, and my son in the surveying and engineering profession, a profession I love.

Chapter 1

ELEMENTARY SCHOOL DAYS

START OF SAMSOG

I WAS NINE YEARS OLD in 1947 when my parents talked about Georgia surveyors forming an association. My mother tried hard to get the name to be GALS—Georgia Association of Land Surveyors. She felt women inspired the men surveyors and wanted the name to reflect such. She also just liked the idea of calling surveyors GALS.

She did not get her wish. The original name in the first constitution and by-laws was the Georgia Association of Professional Land Surveyors (GAPLS). This was changed in a few months to become the Georgia Association of Registered Land Surveyors (GARLS). Mother said this was OK since they really meant GALS, but surveyors did not know how to spell.

The Georgia Association of Registered Land Surveyors (GARLS) became an official organization on June 14, 1947, and was incorporated on May 12, 1949. On July 1, 1964, they changed the name to the Georgia Association of Registered Professional Land Surveyors (GARPLS). Mother said they had the hiccups that day and still had not learned how to spell. The name was later changed, on December 5, 1970, to the Surveying and Mapping Society of Georgia (SAMSOG). That is the name as of the date of this book.

I remember attending the organizational meetings of GARLS in Atlanta, Georgia, with my dad. I sat in the back of the room, reading comic books. Although I do not remember disrupting the meeting, several tell me I did. They said that, during one of the more serious discussions, I laughed so loudly that they could not go on with the meeting. When asked why I was laughing, I held up my comic book.

My dad, R. M. Marbury, Jr., was a charter member of GARLS. In 1954, he became their eighth president.

During June 3–5, 1948, GARLS held a short course in Surveying and Mapping at the Georgia School of Technology. Dad taught a course at this meeting entitled, "Proposed Laws for County Engineers and County Surveyors." Georgia Tech later published the proceedings of this meeting in book form, noting its importance to the surveying profession. Historically, it marked the first Annual Meeting of GARLS. The report noted that this was the first state organization of its kind to be formed in the United States. It was also the first to be affiliated with the American Congress on Surveying and Mapping.

In those days, it was common practice for professional organizations to set up a schedule of minimum fees. The proceedings of the short course on surveying and mapping, dated

June 3-4-5, 1948, suggested minimum fees for miscellaneous survey work. The recommended per diem fee for a registered land surveyor was twenty-five dollars. It was fifteen dollars for a draftsman, ten dollars for an instrument man, and five dollars daily for a rodman and chainman.

The recommended charge for a one-lot survey, reference to a street corner, and platted, was twenty-six dollars. The recommended charge for a one-hundred-acre topographical survey was four hundred thirty dollars.

GARLS recommended a code of minimum basic requirements for surveys and plats. The code required that a compass survey have an error of closure that does not exceed one foot in five hundred feet. A transit stadia survey should have an error of closure that does not exceed one foot in one thousand feet. A transit tape survey should have an error of closure that does not exceed one foot in twenty-five hundred feet.

My father always felt that our surveys should be much more accurate than the minimum standards and insisted that our surveys always have an error of closure that did not exceed one foot in ten thousand feet. This meant that for every ten thousand feet measured by our survey crews, the measurements could not have an error of more than one foot. My dad always insisted that our work meet the highest possible standards for quality and accuracy.

SURVEYING AS A YOUNG LAD

The summer of 1949, at age eleven, I started work as a land surveyor. I would carry wooden stakes, or hubs, for the survey crews. Sometimes I carried iron reinforcing rods. These were used to mark property corners. I carried stakes down city streets, open farmland, and South Georgia swamps. I earned twenty-five cents an hour.

Times were simple in those days, with few modern-day conveniences. Our refrigerator was a compartment that held two fifty-pound blocks of ice. We often purchased the ice from street sellers who rode down the streets in horse-drawn wagons. As horses pulled the wagon near where I lived, my friend Ed Strickland and I would sometimes hop on the back of the wagon and enjoy a short ride.

We had no air conditioning or central heat. In the winter, I would go out to the coal bin in our yard, shovel a bucket-load of coal into an aluminum bucket, and haul the coal into the house. If I could get away with pretending I was asleep, sometimes my dad would do this for me. We then placed the coal in the fireplace on top of kindling and rolled newspapers.

Kindling consisted of what we called "fat-lighter," which was usually small sticks of pine wood with turpentine oozing out of the sides. Dad would strike a match and light the newspapers, which lit the kindling, which, eventually, lit the coal. If we'd remembered to open the damper before lighting the fire, heat from the fireplace quickly warmed the room. If we'd forgotten, the room filled with smoke.

We used fans to keep cool in summer. Sometimes we placed a pan of water or ice in front of the fan to help with the cooling. We always had screens on the windows, and we usually kept the windows open all summer. We did enjoy the luxury of indoor toilets.

ELEMENTARY SCHOOL DAYS

In about a year, I graduated to rear chainman. I held the rear end of a one-hundred-foot steel chain precisely over the point of beginning and over each successive point marked by the front chainman. What we called a "chain" was actually a thin strip of metal one hundred feet long and about one quarter inch wide. Every foot was marked.

We called this an "engineer's chain," as opposed to the earlier surveyor's or Gunter's chain, which was only sixty-six feet long and actually constructed like a chain. I never did any surveying with a Gunter's chain, although I am told that this was the kind of chain used to establish the original city limits of Albany, Georgia, my hometown.

Chaining pins were usually small sticks about one foot long and one quarter inch in diameter. We made them by whittling from other pieces of wood. In later years, we purchased metal chaining pins.

As the front chainman placed the chaining pin at a point one hundred feet away, he would yell, "Come ahead."

I then walked or ran as necessary to reach the placed pin before the front chainman could reach another point.

"Chain," I would yell as the end of the steel chain reached the pin. I held the back end of the chain next to the chaining pin. The front chainman would place another chaining pin at the front end of the chain, yell, "Come ahead," and the process would start over again. In order to keep the chain as straight as possible, the front chainman often pulled so hard on the chain that it was hard for me to hold it securely over the proper point.

GUNTER'S CHAIN AND CHAINING PIN located in the GEORGIA MUSEUM OF SURVEYING AND MAPPING

Soon, I became head chainman. The job meant no increase in pay. What it did mean was that, now, I was the one to drive the stakes into the ground, marking property corners. This job was not so bad on large farm surveys or even when staking subdivision lots. We had to survey several cemeteries during my term as head chainman, however. This meant driving a stake into the ground approximately every six feet. I did not appreciate the extra work then, but it proved a real boon to my future romantic efforts. It gave me strong wrists and the ability to hit small points with precision using a sledgehammer or the back end of an ax.

SUMMER JOBS

I worked on survey crews every summer from the time I was in the fifth grade through my college years. By the time I was a senior in high school, I was a party chief, running all of the survey equipment and making many of the survey calculations.

My dad and granddad both held dual registrations as professional engineers and registered land surveyors. They arrived at the office about seven every morning. They both insisted I, also, get there at seven. While they did office work and made preparations for the survey crews, they gave me survey problems to solve. Before I graduated from high school, I had taken several sample survey exams and passed them all. By the time I took the actual exam, I was among the first to finish and passed with a score close to ninety percent. The strict discipline imposed by my father and grandfather benefited me then and continues to benefit me today.

I became a party chief by default during my high school days. One summer during my senior year, I was doing construction staking for a large sanitary-sewer project. The sewer was between sixteen and twenty feet deep, and our job was to provide cut stakes every fifty feet. Cut stakes were hubs with the depth from the top of the hub to the invert of the pipe clearly marked. One morning, the party chief failed to show up for work. No one could reach him by phone, and no one had any indication why he was not at work. What we did know was that, if we did not keep ahead of the contractor with the cut stakes, the project would shut down, and we could be liable for the damages.

My dad and granddad were busy with other projects, and I was the only one left who had experience running "the gun." In those days, the survey transit was often called "the gun." I also could calculate horizontal closures, vertical curves, and the distance from the top of hub to the required invert elevation of the sanitary sewer. In the past, this was done by the party chief, and I ran a second set of calculations to reduce the likelihood of errors. Now the entire responsibility for the project was mine.

The party-chief position was mine to succeed or fail at. My dad and granddad said, however, that failure was not an option, since we could not afford to pay for any mistakes. I had to do the job correctly and start immediately. Both frightened and excited, I accepted the responsibility. Everyone completed the project on time and on budget. I was thrilled—this further confirmed that the engineering and surveying life was for me.

ELEMENTARY SCHOOL DAYS

My first project was not an example of good construction. As is normal with any sewer sixteen to twenty feet deep, proper compaction was imperative, especially since this sanitary sewer was constructed in the middle of a paved street. The contractors appeared to do a good job of compaction while my dad and I were present to inspect the project. We inspected the project only about once a week, however, and my dad believed the contractors were not compacting the ditch properly when we were away.

Dad suspected this because of how fast construction proceeded. He was once a superintendent for Wright Construction Company and knew how long it would take to lay sanitary sewer in a sixteen-foot-deep trench. He did not believe that a construction company could lay sewer pipe this fast and compact it properly.

To confirm his suspicions, dad found a location about a block away where he could watch the project through the telescope on his transit without the contractor's knowledge. Sure enough, the contractors worked one way when we were present and another when we were absent. Where the specifications required better than ninety-five percent compaction, the procedure used by the contractors would result in less than eighty percent compaction.

After watching for an hour, dad packed up his transit and drove to the site. He confronted the contractors with his observation and required the contractors to remove the fill dirt they had already placed, replace the fill properly, and provide the proper compaction. To be sure they didn't try the same thing again, dad arranged to have a permanent inspector watch the compaction for the remainder of the project.

WATERMELONS IN THE EARLY MORNING

I loved to survey watermelon fields during summertime, especially in the early morning, when the air was cool and watermelons were ripe. There was something enticing to a young lad like me about fresh watermelon.

Each morning, I would look at several watermelons, find the biggest one there, lift it up, drop it to the ground, and watch it explode. Exposed was the red, seedless heart, cool and ready to eat. I would grab the heart with my right hand and cram as much as I could into my mouth. Then I would do the same with another and another, as though I were a monkey sitting in a bed of bananas. The farmers gave me permission to do so, and I took full advantage of their generosity.

I especially liked eating the heart of a watermelon because it had no seeds. Once, during my preschool years, I asked my

mother why a lady walking along the street was so fat. She told me the lady had "pregnitus." She said it came from eating watermelon seeds, and if I didn't want to get the same thing, I had better avoid eating them. To this day, I am careful not to eat watermelon seeds.

HOUSE IN THE MIDDLE OF THE STREET

Surveying is an interesting profession. One survey project involved placing the corners of a proposed home in the middle of a dirt street in south Dougherty County, Georgia. I will omit the name of the client.

The client drove my father to the location. He pointed to a spot in the middle of the street. The street had not yet been constructed but was dedicated to Dougherty County and owned by the County.

"Right there," said the client. "There, in the middle of the street, is where I propose to build my home. You have the dimensions of the house. Place stakes at each proposed corner so the contractor will know exactly where to construct the house."

My father explained that this location was in the middle of a dedicated street. The client insisted my father stake the location of the house at that exact spot. After arguing for almost thirty minutes, my father disgustedly set corner pins precisely at the requested location. Then he left. Dad told me what he'd done and said this was the craziest survey he could remember.

Six months later, my father returned to that same spot. To his astonishment, a house was there, and his client was living in the house. Dad knocked on the door, and the client answered.

"Come in," said the client. "How do you like my new home?"

"You really did it," said dad. "You really built your home in the middle of this dedicated street."

"Yes, and if I live here long enough, and no one objects, the land will eventually be mine."

Ten years later, the client still lived in the same house. He had recorded a plat showing him to be the owner of the land in that location, and, as far as dad and I could tell, he eventually gained legal title to the land. I would not recommend anyone else trying such a stunt, but this client did get away with it. You never can tell about people and the results of their actions.

VALDOSTA DETENTION POND

Many years later, I had a similar situation with a client in Valdosta, Georgia. I had worked with this client for many years, doing both his civil engineering and surveying work. He had a commercial tract of land that he'd developed for a small business. His property was large enough to construct the building and provide the necessary parking but did not have enough additional space to provide for the city-mandated stormwater detention pond.

Adjacent to his property was a small piece of land just the right size for the required pond. The ownership of this property was in question, since neither the City of Valdosta nor any private developer claimed the property. No one was

ELEMENTARY SCHOOL DAYS

sure if the property belonged to the city or one of the surrounding property owners. My client tried various ways to acquire ownership but couldn't succeed because no one could verify who actually owned the property. He asked me if I had any ideas.

I remembered the man who built his home in the middle of a dedicated street. I wasn't going to advise my clients to do anything this foolish, but I had an idea.

"How much risk are you willing to take?" I asked.

"Tell me what you have in mind," he replied.

"Well," I suggested, "you could have me survey the unclaimed property and record the plat. You could then have some friend of yours sell you a quit-claim deed to that property for one dollar. This is perfectly legal, because all a quit-claim deed does is give you the right of ownership based on whatever ownership the seller may or may not have.

"Of course, your friend will have no legal ownership of this site, but you could still record the deed and claim ownership of the property. I am not a lawyer and have no idea if this will or will not work. You asked for ideas, and this is just one of my crazy brainstorming ideas."

My client did just that. I'm not sure who the friend was who gave him the quit-claim deed. It may have been an adjoining property owner who needed space for a detention pond himself. All I know is that my client had me survey the property and record the plat; he recorded the quit-claim deed.

The interesting part of this whole story is that my client proceeded to construct the detention pond on the property that no one had claimed. He had me design the detention pond and submit the plans to the appropriate reviewing authorities; he built the pond.

He did all this with the full knowledge of everyone involved, including the reviewing authorities. Everyone understood the method he'd used for obtaining title to the property. When everything was completed, all involved congratulated my client and me for the innovative way we'd resolved a seemingly unresolvable problem. My client seemed pleased, and for a few months, I walked around scratching my head and saying, "I can't believe it really worked."

MACHINE-GUN TRANSIT

Expect the unexpected. That could be the motto for every surveyor. That certainly fit the situation my dad faced while surveying along a stretch of road in Albany, Georgia. A client hired my dad to prepare a boundary survey along Radium Springs Road. The property was bounded on the east by the highway, and my dad was required to establish the right of way as part of the survey.

In those days, most surveys were done with a transit and a one-hundred-foot steel tape. Dad had a three-man field crew, consisting of himself on the transit and two chainmen doing the measuring.

Dad set up the transit on an existing right-of-way marker, and his crew began the process of measuring the front property line, which was located along the highway. Then it happened.

A shiny new black Cadillac moved southbound in his direction. The chainmen watched the car accelerate. They

ELEMENTARY SCHOOL DAYS

quickly stopped measuring and moved away from the road. The Cadillac continued to accelerate.

Dad grabbed the transit and sprinted away from the road. The Cadillac moved even faster.

Dad ran down the bank, moving about twenty-five feet from the edge of the road. The Cadillac followed. Seconds later, the Cadillac came to a stop, missing dad by a few inches but demolishing the transit.

Surprised and stunned, dad rushed to the Cadillac and opened the door. Inside he found a little old lady, dressed very properly, declaring that she was sorry she missed him but that she was glad that she'd destroyed his machine gun.

"What machine gun?" asked dad.

"The one you had set up on the edge of the highway trying to shoot me in my car."

Dad explained he was only a hard-working surveyor trying to establish the property corners of a small tract of land. What she thought was a machine gun was a transit he used to measure angles.

Embarrassed, the lady explained her behavior. She was wealthy and had endured many attempts on her life. She didn't see very well, and the transit looked to her to be a machine gun pointing in her direction. She was a feisty little lady and wasn't afraid of anything, even some young man trying to shoot her with a machine gun. She apologized and said she hoped my dad would not hold any grudges against her.

Dad said he understood but explained that the lady had just made herself responsible for buying him a new transit. The lady wrote my dad a check covering the cost of a new transit plus some additional funds for the inconvenience. Then, she smiled, climbed back into her black Cadillac, and sped away. Dad later cashed the check, purchased a new transit, went back to the site, and completed the survey.

LET THE WATER FLOW THROUGH

My dad was always unpredictable when doing engineering work for clients who refused to pay. That certainly was true for a client in the Radium Springs area of Albany, Georgia.

The client wanted to purchase property in an area just off Radium Springs Road, where the woods were beautiful, flowers seemed to bloom spontaneously, and wildlife was abundant. There was only one problem. The property was low and subject to flooding.

The client still wanted a complete report, including the approximate level water would rise along the house in the event of a heavy rainfall. Dad completed the report and presented his invoice. His finding showed water would cover most of the first floor of the house. The client angrily replied that he did not believe the report and refused to pay.

Rains came a few years later, and, just as my dad's study had indicated, the bottom floor of the house flooded. The client returned to the office to talk to my dad.

"Alright, you've had your way. Now tell me what to do," grumbled the client.

"First, pay me," said dad.

"Here is your money. Now tell me what to do."

"Gladly," replied dad. "Open your front and back doors so the water can flow through. Then move upstairs until the floodwaters recede."

A HELPING HAND

Sometime in the 1950s, a landowner asked my dad to survey a farm tract in Camilla, Georgia. It seemed like a routine boundary survey. Dad quoted a price and scheduled to start the survey in the next few days. During his standard research, he was surprised to find that everyone he asked about the property expressed concern that he was willing to take the job.

"Why?" asked my dad. "This is simply a routine boundary survey."

"Don't you know?" asked a longtime resident.

"Know what?"

"Don't you know that the last two surveyors who went on that property were shot? One of them is still here in the hospital."

I wonder whether I should still do the survey, thought dad.

He took a little longer to research the history of the property and found that there had been a heated land dispute between his client and the adjacent property owner. The adjacent property owner was the one who had shot the two previous surveyors.

Dad decided he was not going to let an arrogant property owner keep him from doing his job, but he also decided it best to visit the adjacent owner before beginning the survey. As he approached the home of the adjacent owner, he found him sitting on his front porch—shotgun in hand.

"May I talk with you for a moment?" dad asked, keeping a close eye on the shotgun, which the adjacent property owner was now patting with his left hand.

"Come on up."

My dad climbed up the porch stairs and stated carefully that he'd been hired to survey the boundary line between the two properties. As he explained how he planned to go about the survey, an elderly lady stepped out of a car that had stopped in front of the house.

Dad saw that the lady seemed to have problems climbing the stairs and reached out his hand to help. When the lady reached the top of the stairs, the adjacent property owner glared at dad.

"Do you always help ladies up the stairs like that?"

"When I think they need help," dad replied.

"At last I have found a true gentleman," grunted the adjacent property owner, patting his shotgun. "You put that line anywhere you think it should be, and I dare anyone to argue with you."

WORKING WITH MY DAD AS A CHILD

One of the greatest joys of my childhood was working with my dad. I started working with him when I was eleven years old and continued working with him until his death on September 6, 1994.

ELEMENTARY SCHOOL DAYS

I remember getting up at six in the morning to the smell of bacon cooking in the kitchen. Dad would cook breakfast. First, he cooked bacon. Then he would take the bacon from the frying pan and place it on a paper towel, place another paper towel on top of the bacon to absorb the grease, and pour the remaining grease from the frying pan into a small silver pot. The silver pot was a container used to hold the grease until it was needed later.

Dad would crack open two eggs and drop them into the frying pan. Some days he would scramble the eggs, and other days he would prepare the eggs sunny side up. We ate breakfast together in the kitchen while watching squirrels scramble to and fro in our front yard. We lived at 1215 Maryland Dr. in Albany, Georgia. Dad had designed the street and subdivision earlier, and when we first built our home, it was on a dirt street.

As soon as we completed breakfast, we drove to the office. The office was only about one mile away and was a tin warehouse. The back of the warehouse was a pecan-storage area. The front was our surveying and engineering office. Dad and I worked with my grandfather, who was not only an engineer and surveyor but also a farmer. He used the office for both a pecan-storage warehouse and an engineering office.

We generally arrived at the office around seven in the morning. Dad gave me surveying problems to calculate while he scheduled survey field crews and organized for the day. I became so good at survey calculations that, before I'd completed high school, I took a practice land-survey exam and aced it.

Those were good times. Albany, Georgia, back then, was a small town of approximately thirty thousand people. If I wanted to call the office, I would pick up the telephone and listen to the voice on the other end say, "Number please."

I responded by saying, "Number six." That was our office phone number.

Everyone knew practically everyone else in town. Since the operator usually recognized my voice, sometimes I would just say, "I want to speak to my dad." The operator rang the office, and soon I was talking to my dad. Times sure have changed.

Chapter 2

HIGH SCHOOL DAYS

LOVIC CARL MARBURY

FOR THE MOST PART, my high school days were happy times. School started the day after Labor Day and was usually out near the end of May or first of June. During the months of June, July, and August, I worked with the survey crews.

Around the beginning of my sophomore year in high school, my mother advised us that I was having a brother—or maybe a sister. I grew up as an only child and had always wanted a brother or sister. On February 12, 1954, my brother, Lovic Carl Marbury, was born. He was named after my mother's brother Carl and my dad's brother Lovic, who was a fighter pilot during World War II and died during the war. He is remembered every Memorial Day in Albany, Georgia, by a cross set up in a field of flags honoring those who gave their lives during that war.

The birth, however, was difficult for my mother. She was in tremendous pain, and the birth took many hours to complete. Medicine was not as advanced as it became years later, and because my mother was diabetic, she could take almost no pain medication. Not only did my mother experience extreme suffering during the birth experience, but my brother died the next day, on February 13, 1954. The doctors said the death was due to the prolonged delivery.

Dad and I were heartbroken. Since my mother was not well enough to attend the funeral, dad and I did not attend the funeral, either. We just felt that our place was with my mother. All of us grieved over the death of my brother for many years. Mother was never able to have another child, and, as a result, I grew up having no brothers and no sisters. I was fortunate, however, that both my mother and dad spent many hours with me. As a result, our family was always close.

ROMANTIC BENEFITS OF SURVEYING

I learned the romantic benefits of surveying on a trip to Jacksonville Beach, Florida, in May of 1955, during my junior year in high school. It was my first date with a beautiful band majorette named Fonda Starnes. Our Albany High School band was performing in Jacksonville, Florida. Fonda and I were both in the band, and, during the trip, the band visited a park at the beach. Besides the rides and shows, there was a booth giving prizes to anyone who could hit a round object hard enough to ring a bell about twenty feet in the air.

The booth was similar to those featured in strong-man comics. It looked like a six-inch-wide board about twenty feet high with a wire running from the ground to the top. Attached to the wire was a metal ball. A small round bell sat atop the board.

A bar about six feet long rested on the ground, acting like a seesaw, with the metal ball sitting on top at one end and a round knob at the other. The object was to hit the round knob hard enough that the metal ball would bounce to the top of the board and ring the bell. No one had been able to ring the bell during the last twenty or thirty minutes.

Fonda and I watched as person after person tried in vain to ring the bell. Football players tried and failed. Bodybuilders tried and failed. Weightlifters tried and failed. Graduated markings indicated scores from one to ten. If you could hit the knob hard enough to register a ten, you would ring the bell. Most scored six or seven. One football player scored an eight, but no one could ring the bell.

Then, in an attempt to embarrass me, one of the football players said I should try. I weighed only 135 pounds at the time and had no desire to embarrass myself by scoring only a four or a five, but everyone at the beach kept goading me to try to ring the bell.

Ding! I rang the bell on my first attempt. *Ding!* I rang the bell again. *Ding!* I rang the bell for a third straight time and won a teddy bear for Fonda. Those months of driving stakes in the ground while surveying cemeteries and subdivision lots paid off. Fonda and I married seven years later and recently celebrated our fifty-seventh wedding anniversary.

FONDA AND THE SANITARY SEWER

Surveying in the middle of the twentieth century was different from what it is in the twenty-first century today. We had no distance meters, no GPS (Global Positioning System), and no mobile telephones. We had few safety requirements, and those we did have, we usually ignored. I climbed 150-foot-high water tanks with no safety harnesses, scrambled to the bottom of live sanitary sewers to measure invert and pipe sizes with no safety devices, and waded in swamps filled with water moccasins and rattlesnakes without snake boots or waders.

Sometimes we worked all morning without water, returning to the work truck at noon to quench our thirst and devour a brief lunch. The workday started at sunup and ended at sundown. Life was good; I was young, and I loved it.

Although, as I stated earlier, the stamina and strength from the exercise and physical exertion of surveying had its romantic benefits, there were also times that survey projects were anything but romantically beneficial. The job of measuring the location and inverts of sanitary sewers around Albany, Georgia, was one of those projects.

My responsibility was to climb down each sanitary manhole, measure the depth to the sewer invert, and measure the diameter of the sewer. Temperatures that summer ranged from ninety to more than one hundred degrees Fahrenheit. Live sewage flowed in many of the manholes, and I took care to avoid having sewage douse me in the face while measuring. As I worked, sweat drained from every pore of my body, and, sometimes, it was a toss-up as to what smelled worse—me or the sewage.

Near the end of a tedious workday, I climbed into a manhole close to the home of Fonda, my future bride. We were in high school then; we had been on very few dates, and I tried every day to impress her any way I could. That day, I completed my measurements and climbed out of the manhole.

"Hi, Fonda," I said, peeking my head up from inside the manhole, looking like a roach-infested vagabond and smelling like rotten eggs.

"How about a date tonight?"

Peering out of the manhole as I crawled onto the street, I looked up to see Fonda staring in disgust at my dirty, soiled, stinky carcass. I impressed her alright—but not the way I'd planned. No way was she willing to go on a date with me that night.

THREE MEETINGS

Sometimes people get so caught up in meetings that they fail to do anything but meet. This was the case in my early youth.

Dad and I had the opportunity to design a sanitary sewer lift station located a few feet from a large dike. The dike served to protect the surrounding property against flooding from the adjacent river. As construction neared completion, dad observed a defective valve in the drainage pipe going through the dike. Fixing the valve was a simple matter, and he reported it to the maintenance department.

If the river rose above a certain elevation, and the valve was not repaired, the property would flood. This would result in floodwaters reaching the pumping station and causing thousands of dollars worth of damage.

About six months later, the repair had not yet been done. Again, dad reported the need to fix the valve, stressing the low cost of repair and the high cost of potential flood damage. Several more times he stressed the need to do something. Each time those in charge of maintenance listened attentively but did nothing.

In time, the floods came. The valve had not been repaired, and thousands of dollars was lost in damages. Frustrated, dad approached the maintenance department.

"Why didn't you do something?"

"We did."

"Well, what did you do?"

"We had three meetings!"

DON'T MESS WITH A DEAD ALLIGATOR

The outdoor life attracts many to join the survey profession. We see terrain that others never see as we traverse open fields, well-manicured plantations, cultivated farmlands, and swamps. Wildlife skips, flies, and dances around us as we observe Mother Nature at her finest.

We also learn to watch out for the dangers in the outdoors such as snakes, grizzly bears, and alligators. Seasoned surveyors avoid these dangers whenever possible. Those new to the profession sometimes act foolishly until they learn better. This is an account of one of those foolish acts.

It was a balmy summer day in late July. The temperature was approaching one hundred degrees, and our job was to survey a large tract of swampland in South Georgia. I was just out of high school and about to begin my freshman year at Georgia Tech. My dad let Tommy Herrington, a good friend, and me work for his surveying company during the summer. Dad was the registered surveyor. Tommy and I were his crew.

We left early on a Monday morning and arrived at the swamp about 8:15 a.m. Even though it was wet, we wore boots less than knee high and no hip waders. It was just too hot for hip waders. Dad carried the transit. I carried the chain and chaining pins. Tommy carried an eight-foot-high range rod. That day, the survey began near the edge of a small creek.

We maneuvered our way through the swamp, reaching the creek in five or ten minutes. Tommy carried the rod to a point set the day before and placed the rod on the point. That was to be the starting point for the survey. Then it happened.

Just off the creek bank rested a large alligator. It looked dead. It was not moving, and its eyes were shut. It must have been about ten or twelve feet long. I yelled for Tommy to get away. He yelled back that the alligator was dead. Dad yelled again for Tommy to move. We would find another place to work until the alligator moved. Again, Tommy yelled that the alligator was dead.

Determined to prove a point, Tommy took the range rod and poked the alligator in the eye. The dead came to life. One snappy flip of his tail, and the alligator had the rod sailing through the air like a javelin. The alligator looked at Tommy. We all looked at each other. Then, in what seemed like only a few seconds, we were all safely back at the survey truck.

HIGH SCHOOL DAYS

I have heard that no one can run one hundred yards in less than 9 seconds, but I believe we all broke the record that day. We were all so nervous—actually, scared—that my dad worked with two other crew members to finish the job.

HILLSMAN PARK

One of the early projects dad and I worked on was the design and construction of Lake Loretta. Spencer Walden, a local developer and close friend, envisioned a large lake near the Dawson Road in Albany, Georgia, with homes surrounding the lake. Dad designed the lake, I helped with the survey, and the project was an immediate success.

The success of the construction of Lake Loretta and several other lakes in the area prompted the City of Albany to construct a lake just north of Third Avenue, which they called Hillsman Lake. Both my father and grandfather advised the city not to construct the lake because it was in a limesink area. Over my father's and grandfather's objections, the city proceeded to design and construct the lake.

The lake covered about a city block and was five or six feet deep. The city used a large pump to provide enough water to fill the lake.

One day my father noticed that sand as well as water was spewing from the pump. Taking a closer look, my father believed that the sand was coming from underground limestone caverns. He believed that if the pumping continued and the lake continued to fill with water, a limesink would result. Concerned about the safety of the residents and anyone who might venture into the lake, my father went to see the city engineer and city manager.

After a heated discussion, the city manager responded, "You say this lake is going to cave in. If you're so smart, just when will it happen?"

Angrily, dad replied, "Probably about six or seven o'clock tomorrow morning."

What no one in the room knew at the time was that a reporter from the local newspaper was listening to the conversation. As fate would have it, a large limesink developed precisely at the time my father had predicted. In a matter of hours, the lake was dry.

The story made the front page in the local newspaper, creating anxiety throughout the city. My grandfather quickly drew an illustration explaining precisely what had happened and why this was not a threat to the surrounding neighborhood. That illustration was in the following day's newspaper.

I questioned my father regarding how he was so sure that the lake would fail. First, he said he knew that the city had constructed the lake in an area with known underground limestone caverns. He then explained that the dirt pumped out of the well was identical to the dirt at the bottom of the lake. That meant that water was recirculating from the water in the lake through the ground and back out of the pump. That also meant that a limesink was forming and that a cave-in was imminent.

Although construction of the lake ceased immediately, the area is now a significant recreation area within the city.

HIGH SCHOOL DAYS

The city changed the name, however, from Hillsman Lake to Hillsman Park.

SPIRAL STREETS

I worked at Marbury Engineering every summer during my high school and college days. Civil engineering and land surveying seemed to be part of the same profession in the late 1950s, and I loved doing both.

We did most calculations by hand in those days, and often the calculations were tedious. It was not unusual for subdivision calculations to fill twenty or thirty pages. Competition was fierce, and my dad thought of a way to keep survey competitors out of subdivisions we designed.

"How many surveyors in this area do you think know how to calculate spiral curves?" dad asked.

"Not many," I replied. "They probably all learned in school, but they probably do it so seldom that most have forgotten."

"Well, we are not going to use circular curves in this new subdivision. Every curve will be a spiral. Either our competition will not know how to set lot corners along a spiral curve, or it will be so much work that they will forget about trying to do work in subdivisions we design. It will be a little more work for us in the initial design, but we will have a lot of fun watching our competition try to figure out how to place lot corners around a spiral curve."

My dad did what he promised, and we designed all curves in the new subdivision as spirals. It worked better—or perhaps worse—than planned. Other surveyors did not know how to duplicate where we located the lot corners. As a result, most placed survey monuments in the wrong locations, and, later, we had to go back and correct most of them.

Our plan did not keep our competition away. It just kept them from placing lot corners in the right location. We seldom got paid for resetting the corners. That was the last time we did such a crazy thing, but we did have fun doing it, and we had fun watching our competition grumble while trying to place lot corners in such illogical locations.

STREET NAMES

I enjoyed naming streets. My father, grandfather, and I did a lot of that as we designed subdivision developments. Sometimes we named streets for friends and relatives. Sometimes we named streets based on subdivision themes, and sometimes we named streets based on a play on words. Naming streets based on a play on words often brought fun memories to our entire staff.

One street caused several of us to come home broken out with rashes over much of our bodies. We cut paths down the proposed middle of the street, grabbing limbs and bushes with our hands and throwing them to one side. Sweat streamed from every pore, and we wiped the sweat from our faces with our forearms as we continued to work. The job of measuring one street took several days, and we could hardly work due to the intense itching over our face, arms, and feet. We named the street Poison Ivy Lane.

That name did not work well for the owners, who had to market the development and sell lots along the street. We had to change the name to Ivy Lane on the recorded plat, but, to us, the name is still Poison Ivy Lane.

John Herndon surveyed many of the streets in a subdivision later named Sherwood Acres. We named all the streets in that development after characters or locations found in the stories of Robin Hood. We were not able to name any of the streets John Herndon Lane, but we did name one Little John Lane. John Herndon knew that this was in his honor and seemed pleased.

Someone named a street Sewage Treatment Plant Way, since this street led to the local sewage treatment plant. We recommended that the County change that street name. They did so promptly.

The Dozier family owned and developed Merry Acres Subdivision in Albany, Georgia. They named one of the streets Marbury Lane after my grandfather. We all considered this an honor and appreciated the friendship our family had with the Dozier family for so many years. None of our family ever lived on this street, although I once strongly considered building my home there. It was and still is a beautiful street. It would be an honor to live in that neighborhood. My wife and I felt we might appear too pretentious if we lived on a street with our name on it, however, so we decided to build elsewhere. I still sometimes wonder if we made the right decision.

Nancy Cartmell served as our secretary for many years. One year, we designed a street going north from Ledo Road into Lee County, Georgia. The street was adjacent to an expensive commercial development. We felt that Nancy was a valuable asset to our company and wanted to name a street after her. In some way, the value of the land fronting on the street would symbolize the value we felt Nancy had to our company and the community. We named the street Cartmell Lane.

Of course, the developers had to approve all name recommendations, and local street-naming committees or planning commissions also had to approve street names. Still, our company named many streets and roads in developments we designed, and we enjoyed being part of the process.

SURE CURE FOR ASTHMA

We appreciate all our clients, but some are much too nosy. This presents two problems. One is the problem of maintaining a client's confidential information while another client insists on wandering about the office. The other problem is trying to work while the client constantly interrupts, talks, and moves things that should be left alone.

My father had one of those nosy clients on a summer afternoon in the late 1950s. The client was pleasant enough but a little too curious about how we made reproductions of survey drawings. He walked around the office handling everything in sight and moving papers from one desk to another without a thought as to the impact on our production. Among his many questions, he wanted to know exactly how we made blueline drawings.

We usually made original drawings on vellum tracing paper. This smooth, durable paper had high transparency, allowing light to pass through. We drew on this type of paper in order to produce exact copies of our completed final drawings.

HIGH SCHOOL DAYS

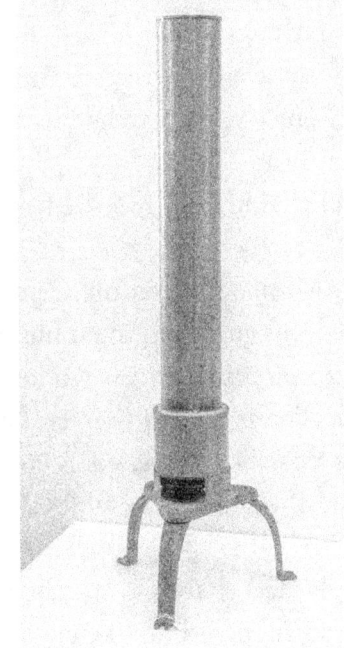

OLD BLUEPRINT MACHINE AND AMMONIA TUBE USED FOR DEVELOPING

Drawing reproduction was not as simple as it is today. We reproduced drawings using a blueprint machine at the far end of our office. The machine was actually just a round glass cylinder over which we laid the original drawing. Inside the glass cylinder were high-intensity light bulbs. We laid a sheet of yellow-coated paper over the original drawing, wrapped a cloth cover over both sheets of paper, and turned on a timer. This caused the lights inside the glass cylinder to glow for two or three minutes.

When the lights went off, we removed the yellow-coated paper from the glass cylinder and placed it in a vertical metal cylinder. A small cup of ammonia rested beneath the metal cylinder. The ammonia fumes drifted up from the cup into the metal cylinder, causing the yellow-coated paper to develop into a blueline print of the original drawing.

Our client watched as the lights went on in the round glass cylinder. When the lights went off, he watched as we placed the yellow-coated paper into the cylinder.

"What is that vertical metal cylinder for?" asked our nosy client.

"That is a sure cure for the asthma," replied my father. Dad was too tired of our client's many questions to go into a long explanation of how the ammonia print process worked.

Without another sound, our client grabbed the vertical metal cylinder, held it up to his nose, and inhaled deeply. One second later, he fell across one of our drawing tables and onto the floor.

Dad rushed to help and explained that he'd just been joking about the cure for asthma. Our client took the experience good-naturedly.

"My asthma is gone," said our client, "but that sure is a hard way to cure it."

We continued working for this client for many years, but he never meddled with our office supplies again.

WATER ON LINEN

In the late 1940s and early 1950s, we prepared some drawings on vellum tracing paper and other drawings on transparent linen, or drafting linen. This was a piece of linen cloth, prepared in a way that allowed us to draw on one side of the medium using India ink. India ink was a permanent ink with archival qualities. Once, my grandfather drew a map of the plantations of Southwest Georgia on drafting linen. As early as June 8, 1926, my grandfather prepared the original official map of the City of Albany, Georgia, on drafting linen.

Drafting on linen produced beautiful finished products, capable of easy reproduction. The product had one major flaw: Water made permanent white splotches when the linen got wet.

My grandfather was near completion of a major drawing when disaster struck. The front part of our office was for engineering and surveying. The back was a pecan warehouse. All the floors were wooden, and sometimes dust from the pecan-warehouse part of the office found its way into the engineering section. At those times, someone would sprinkle water on the wooden floor to reduce the effect of the dust. Unfortunately, that was one of those times.

In an effort to reduce dust, a well-meaning staff member sprinkled water about the room. About three drops fell on my grandfather's almost-completed drawing. I was in my early

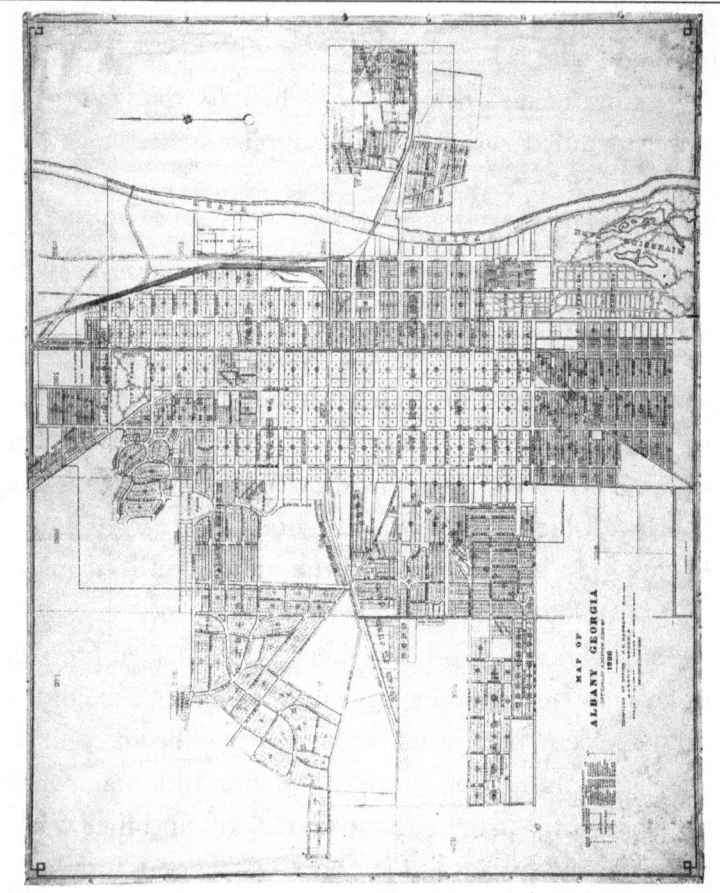

**ORIGINAL MAP OF THE
CITY OF ALBANY, GEORGIA
COMPILED AT THE OFFICE OF
R. M. MARBURY, SR.
MY GRANDFATHER
JUNE 8, 1926**

teens and had a relatively good vocabulary. My grandfather blurted out some words, however, that I had never heard. If words were explosives, his words would have obliterated the entire building.

Everyone evacuated the premises within sixty seconds. None of us came back until the next day. I don't know if my grandfather redrew the plat or if he completed it allowing the white spots to remain. I didn't ask, and no one else did, either.

SURVEYING RIVERS AND CREEKS

Often, we received contracts to survey riverbanks or lots along creeks. Sometimes the lot lines were along the shoreline, but often the lot descriptions called for the rear lot lines to be the center of the creek. This presented both challenges and opportunities. The challenge was measuring the creekbed. It entailed avoiding mosquitos, gnats, and alligators—and not drowning in deep water.

Measuring the creekbed was also an opportunity. You see, this meant we had to purchase a boat. Surveying from a boat meant we could do a little fishing during lunch.

My first experience in river surveying was measuring the Flint River in Albany, Georgia. I was in high school, and a favorite pastime was fishing with my dad. We had a lump-sum contract, so our clients paid us the same amount regardless of the time it took to complete the survey. My dad purchased a boat and a small motor for the company. Then we went to work.

We established a baseline along the west bank of the river. We placed wooden hubs at one-hundred- or two-hundred-foot intervals along the bank. We noted the location of each hub on a map. As we worked, we also noted good fishing locations. We worked and fished; then we worked and fished some more.

Once we'd established the baselines, dad and I climbed into the boat, carrying with us a twelve-foot-long level rod. The level rod was a twelve-foot-long pole with one-foot graduations up the side. Between each foot mark were graduations for each one tenth and one hundredth of a foot.

We drove the boat to a point at the center of the river, anchored the boat as best we could, and measured the depth of the water with the level rod. We then held the level rod vertically in the air.

An instrument operator set up a transit over each hub and measured the distance to the level rod, using the stadia hairs on the transit. He would measure the distance between the stadia hairs in tenths of feet and multiply that distance by one hundred.

If the distance between the stadia hairs was 1.54 feet, we knew we were located 154 feet from the instrument. We would also measure the angle from a known point to the level rod. Knowing the distance to the level rod—and the angle from a known point—we could calculate our exact location on the river. This was a slow process, but it was the best way we knew how to do the job at the time.

Near the end of the day, we gathered our survey equipment, placed it in the truck, and took the boat back into the river. This time, we took with us large pond worms and fishing tackle. I don't remember if we caught many fish, but I do remember the good times I had fishing with my dad.

HIGH SCHOOL DAYS

Years later, we found better methods of measuring rivers. We purchased total stations with the capability of measuring distances by looking at glass prisms. We would set up a total station at some point of known location, take a glass prism in the boat, and travel up and down the river. The instrument operator would measure the distance from his location to ours and record the distance and angle. We plotted our location on maps after we returned to the office. We didn't have much time for fishing during those days, but we did use the boat we purchased to fish on weekends.

As technology continued to improve, so did our ability to measure the locations of rivers and creeks more efficiently. With the advent of GPS (Global Positioning System) technology, we were able to take GPS equipment with us on the rivers and creeks and determine our location as we drove our boat along the river. On one project, we used aerial photography to measure locations along rivers and creeks. We surveyed to each end of the river noting the location. We then noted these locations on aerial photographs drawn to a known scale. From this information, we plotted the river location from scaled distances on the aerial photographs. This was not a very precise method, but it was good enough for the required accuracy of that specific project.

Another enjoyable way of measuring rivers and creeks was by airplane. Both dad and I were pilots. With GPS, we could put a GPS unit in the airplane and fly just over the top of the river, letting the GPS monitor our location as we flew. We never actually used this procedure for river-location surveys, but a good friend and former business partner, Jody Logan, did.

41

Jody was a pilot and told me that he'd done a survey of many miles of riverbed for a fraction of the cost quoted by other surveyors. He did the survey using his airplane and a GPS unit. When he told me what he'd done, I searched for a similar project to survey by airplane. I never found such a project, but I always thought that would be a fun way to survey a river.

SURVEYOR SPANKS CONTRACTOR

A young surveyor worked for my father during his younger years. He was a big man, tall, and muscular.

My dad tells the story of this surveyor and a local contractor. The surveyor was party chief on a survey crew at the time. His job was to measure the amount of sanitary sewer constructed in order to determine how much to pay the contractor for his work. Contractors were paid based on the amount of sewer line constructed and based on the depth of the sewer. The deeper the sewer, the more the contractor was paid.

Sometimes the depth of the sewer was between one pay item and another. That is, sometimes the pay item was a set amount for a depth between six and eight feet and a greater amount for a depth between eight and ten feet. Sometimes the measurement would fall at exactly eight feet. If the measurement showed a fraction of a foot deeper than eight feet, the contractor received the larger amount. If the measurement showed a fraction of a foot less than eight feet, the contractor received the smaller amount.

The contractor asked the surveyor to always show the depth on the side where the contractor would be paid the most money.

Dad was in his office shortly after the contractor made that statement to the surveyor. He did not know of the conversation between the surveyor and the contractor at the time.

The surveyor walked into the office dragging the contractor by his collar. He pushed the contractor into a private room and closed the door. The next thing my dad heard was a loud slap, and then another, and then another. Dad opened the door to see what was happening.

The surveyor had the contractor across his knee and was spanking him like a baby.

"He tried to bribe me," said the surveyor.

The contractor had probably made those statements in jest, but I don't believe he ever did that again.

THE SURVEYOR AND THE SKULL

My dad had another interesting experience with a young surveyor. One day they were working in a new subdivision. It was hot, the woods were thick, and the field crew had to use bush axes and machetes to cut their way through most of the brush. On previous occasions, near that same area, the crew had discovered a moonshine still, and on other occasions, the crew spent much of their time fighting rattlesnakes and water moccasins. This day was also one of those unforgettable days.

The surveyor looked at the ground. He saw an unusual white object grinning up at him. Closer inspection revealed that this was no ordinary white object. It was a human skull. He picked it up and told my dad he wanted to take it home and place it on his mantle. Dad explained the seriousness of this idea and insisted that he take the skull to the local police. He did.

Later, the police determined that this was the skull of a man missing for many months. Apparently, he'd been bitten months before by a rattlesnake or water moccasin. He died there in the woods. You can never tell what you will find while surveying in the swamps.

DAD HELPS SICK COMPETITOR

In my early days of surveying and engineering, competition was not as "cutthroat" as it sometimes is today. We competed, yes, but usually on a friendly basis. Some of our closest friends were John Sperry, Jack Holland, Bill Lowe, Jack Dean, Donnie Lanier, and others, all of whom were surveyors and engineers. Ben DeVane, Stan Folsom, Dent Reeves, Wiley Rice, Bryson Langford, and Buddy McCoy were among other close friends and competitors in the surveying and engineering business.

My dad's experience with Jack Holland is an example of how we all worked together. Jack was a big man with a deep voice and large presence. He marketed well and always had his share of engineering work. On one occasion, he became sick and had to enter the hospital for an extended period. He had considerable work under contract, and failure to complete the work would place him in a serious financial position.

Dad met with Jack's staff and obtained a list of his ongoing projects. Dad devoted his staff, time, and energy to completing Jack's projects one by one, so that Jack would not default on any of his contracts. Jack's clients never knew my dad did the work, and all of the income from the work went to Jack

HIGH SCHOOL DAYS

and his staff. That was the way it was in those days. We all helped each other.

Dad did have some fun with Jack after Jack's recovery. Dad told Jack that his quality of work had really improved while he was in the hospital. I never heard Jack's response, but it was probably something like, "Yes, and I knew I had better get well quick before my clients wanted me to be in the hospital all the time."

Jack was always a good sport and a good friend.

INTERESTING CHARACTERS

I met some of the most interesting characters alive during my surveying days. One was an eccentric doctor I met while surveying the site of a fast-food restaurant. I will not give his name, since, at the time of this writing, some of his family still live here in Albany, Georgia. The doctor was a kind old man who loved to see people's reactions to his eccentric behavior. My survey crews talked about the hard time they were having due to financial challenges.

"Money is not a problem," said the doctor. "It is really good only for toilet paper."

With that said, the doctor reached in his pocket, pulled out his wallet, grabbed a wad of dollar bills, and tossed them all over the yard. Worked stopped, as my survey-crew members ran back and forth, picking up dollar bills. Every time they started back to work, the doctor would toss more money into the yard. Sometimes he tossed one-dollar bills, sometimes five-dollar bills, sometimes even ten-dollar bills. We did not get much work done that day, but as best I can recall, we picked up more than one hundred dollars tossed away by the eccentric doctor.

Another client had a car that floated on water. He was a wealthy client from out of town and enjoyed doing the unusual. Our company designed a mobile-home park for him, and, near the end of the project, he told me that he wanted to have some fun with the people in our town.

The Flint River flows along the middle of town, and a bridge crosses the river at Broad Avenue. During one day of heavy traffic, the client drove his car to the bridge and then veered off into the water. He drove his car straight into the Flint River, not even slowing down. Crowds rushed to the scene, and some even dove into the water, fearing that the car would sink and the driver would drown. Our client cheerfully continued driving his car across the river and up the bank to the other side of the river. Some people are just naturally jokesters.

We also had our share of unique staff members. One member of our survey field crew came into the office one day asking my dad to help him with a lawsuit for child support. A young woman claimed the worker had gotten her pregnant and wanted him to pay expenses. The worker did not want to pay.

At court, the judge told the worker he never should have been involved in such an act that resulted in this woman getting pregnant. Our worker looked the judge straight in the eye and responded, "But, judge, I said 'excuse me.'"

On another occasion, one of our hardest workers came to our office on a Saturday morning crying and asking my dad for help.

HIGH SCHOOL DAYS

"I think I just killed a man," the worker told dad.

"Are you sure?" asked dad.

"I think so," said the worker. "Both of us were drunk, and he told me he was going to kill me. I told him he would not kill me because I would kill him first. He told me again he was going to kill me, and I told him again I would kill him. When he again told me that he was going to kill me, I told him that I would show him I would kill him. I went into my house, got my shotgun, and shot him."

"Did it kill him?" asked dad.

"I think so," said the worker. "There was his head. Bang. Weren't no head."

The worker then started crying. He kept asking, "What will my mother think?"

Dad called the police and took the worker to the police station, where they put him in jail for murder.

After serving several years for murder, he was released and came back to work for us. He worked hard and was one of our best workers. No one condones murder, and this was a terrible event. Nevertheless, he worked hard after getting out of jail and never caused any trouble again, as far as I know.

Chapter 3

COLLEGE AND MILITARY

ALBANY SHOPPING CENTER— MIDTOWN MALL

In 1959, Spencer Walden moved the first shovel of dirt at the site of the Albany Shopping Center, later called Midtown Mall. My father did the site engineering for the project. To my knowledge, this was the first major shopping center for the City of Albany, Georgia. Stores planned included Western Auto, WT Grant, Winn-Dixie, Woolworth, Midtown Drug Company, and several others. The shopping center would have free parking for approximately one thousand cars.

Site construction for the shopping center did not always go smoothly. One day, as my father inspected the work, he noticed major deficiencies and approached the contractor.

"Do you think you're going to complete this paving project today?" my father asked.

"If that short-legged SOB doesn't catch us first," the contractor replied.

"Well, you're looking at him," replied my father. "Remove all of the faulty work, and redo it." My father then walked away.

Sometime later, my father had another experience with the same contractor. While inspecting the site, two of the contractors approached him from the rear. One of the contractors hit him over the head with a monkey wrench. Dad was wearing a leather cap, which somewhat softened the blow. Having unusually quick reflexes, dad managed to slam his foot against the insole of the first man's ankle, breaking it. He also managed to hit the flexible knee of the second contractor, breaking that contractor's leg. Later, ambulances took all to the hospital.

My father was a once a Golden Gloves boxing champion and also a martial-arts expert. He served with the U.S. Navy Seabees in World War II and was involved in combat operations in Saipan, Tinian, Iwo Jima, and other operations in the Pacific. Much of his fighting during World War II involved hand-to-hand combat. Dad never liked to talk about his combat experience, but no doubt, this training and experience served him well when he was attacked by those unethical contractors.

Dad suffered a concussion from that experience, which may have contributed to his suffering from Alzheimer's during the last years of his life. We never had trouble with those contractors in the future, however.

That year—1959—was a year I remember well. It was the year before I graduated from college. The maximum Social Security check for a family that year was $254 per month.

Gasoline was $0.25 per gallon, and Hawaii and Alaska became the 49th and 50th states.

DARTON STATE COLLEGE

During the early 1960s, the Georgia Board of Regents hired Collier Houston as the architect of a proposed junior college in Albany, Georgia. Marbury Engineering Company was selected as the surveying and civil engineering company to do the site surveying and design. We did an engineering study, and the Board of Regents selected a site on the south side of Gillionville Road in Albany, Georgia. The topographical survey of this site proved interesting.

I was the party chief of the surveying crew doing the topographical survey. Our crew performed the survey using a plane table with a telescopic alidade and a Beamon arc. The plane table was actually a wooden board about four feet long by four feet wide that mounted on a tripod. The telescopic alidade was a twenty-power telescope with stadia hairs located inside the telescope. A rodman would hold the stadia board at various locations along the ground, and, as the instrument man, I could read both the elevation of the ground and how far away the point was that I was observing.

The alidade had a straight edge on the bottom, and, using that, I was able to plot the exact location of each point as well as its elevation. I drew the survey on the plane table while making measurements on the ground. When I returned to the office, I would trace the survey onto another sheet and make a more detailed drawing to give to our clients. Our survey crew averaged surveying more than forty acres each day.

51

PLANE TABLE WITH TELESCOPIC ALIDADE AND BEAMON ARC

COLLEGE AND MILITARY

The stadia board was a fifteen-foot-high board about six inches wide. It was like a large level rod in that distances were shown in feet and tenths of a foot. I could read the distance between the top crosshair and the bottom crosshair, multiply that distance by one hundred, and that would give me the distance from the stadia board to my instrument.

It folded in the middle so that the rodman would have to carry a board only about eight feet long when going to the first point. We used to call it the "finger-crusher," since sometimes the stadia board would crush the rodman's finger if he forgot to move his finger when folding the rod.

The site selected was an area where rattlesnakes traveled on a regular basis. During the two or three months required to do the topographic survey, my crew and I killed an average of two poisonous snakes every day. I was young then and actually enjoyed bringing home rattles from the rattlesnakes I had killed.

Some of the other crew members would skin the rattlesnakes after we killed them and use the skins to make belts. I am older now, and today I would work much slower and carefully in an area that was so snake infested. In those days, we didn't even wear snake boots. We really did not act very bright.

After completing the topographic survey, I prepared the final drawings. My father did most of the site design on the project. I remember that a great deal of fill material had to be placed on the site. When the project was bid, the successful contractor quoted a price of twenty-five cents per cubic yard to place fill material on the site. Even then, that was a low price. The contractor explained that he was able to obtain fill material from a close location at practically no cost to his company, and that was how he was able to give such a competitive price.

Darton State College was originally called "A College" on the first set of plans. The developers held a contest to determine the name of the college. The winning name was Albany Junior College. I always wondered what other names were suggested that allowed the winning name to be Albany Junior College. That name just never sounded very original or unique to me. Later the name was changed to Darton College and then to Darton State College.

The first president of Albany Junior College was Bill Tilley. He was an outstanding president and did much for the college. I remember two interesting things about him. The first was that he had a real interest in plants, and second was that he was opposed to any significant sports program in the early years of the college. In later years, when Albany Junior College became Darton College, sports programs were added, and Darton College became well-known for its sports programs. They even went on to win national championships in sports such as golf.

In 2017, Darton State College merged with Albany State University and became the west campus of Albany State University.

LAKE PARK SUBDIVISION

During my time in high school and college, many developers in the Albany, Georgia, area were developing subdivisions

COLLEGE AND MILITARY

around lakes. One of the major developments my dad and I worked on was Lake Park Subdivision.

Surveying the area later known as Lake Park Subdivision brought many unexpected experiences. Most notable were the findings during our surveys. I wrote earlier of experiences with a young surveyor and a human skull. Another experience was the equipment I found when surveying an area only a few hundred yards from where I eventually made my home.

In a small patch of woods, hidden from view, was a steel drum with a flexible pipe going from it to a smaller container. It also appeared that someone often built a fire around the larger container. I had no idea what I had found. Dad laughed when he saw it. It was a moonshine still. They never caught the moonshiners, but the law-enforcement officers destroyed the still.

Lake Park subdivision was originally designed with two lakes. One was named Lake Cornelia and the other Lake Loretta. The lakes were named for daughters of the Haley family, one of the prominent families in our city.

Dad designed the lakes with pumps supplying the water necessary to fill them. He designed one pump for both lakes, with the water from Lake Loretta flowing across a spillway, through some pipes, and into Lake Cornelia. The lakes did not get filled, however. Water percolated out of both lakes faster than the pumps could meet the demand. Also, a limesink developed in the middle of Lake Loretta.

After considerable research and conversation with the developer, Spencer Walden, they decided to focus on keeping water in Lake Loretta and allowing Lake Cornelia to function as a large park. The problem remained as to how to seal the limesink in Lake Loretta.

Dad recommended using a substance called bentonite. This is a type of clay, mined and sold by a company in Alabama, that expands many times its volume when wet. Dad and Spencer decided to create an island in the middle of Lake Loretta at the limesink location. The island would do two things. First, the island would be above the water level of the lake and as such would prevent additional water from seeping into the limesink. Second, the island would provide an additional aesthetic feature to an already beautiful lake. It worked. The island also became a bird sanctuary. Various types of birds, from ducks to eagles, visit the island every year.

One of the streets in Lake Park Subdivision is a winding street called Lullwater Drive. Dad originally designed this as a straight street connecting the Dawson Road with Gillionville Road. Spencer did not like the street being straight and told my dad so. Like so many surveyors and engineers, dad wanted the street to provide an easy connection between the two destination points and felt his design was best. Spencer felt a curved street would provide a more pleasing location for future homes. After considerable discussion, dad told Spencer to show him where he wanted the street to go, and dad would put it there.

Spencer placed markers in the ground in random locations requiring considerable curves. Dad designed the street accordingly. The winding street was a good decision and added beauty and more aesthetic appeal to the area. The numerous curves also served to slow traffic as it moved from one major street to another. To be honest about the slow traffic, however, an abundance of traffic tickets given to speeders along this route was really the major contributor to the slower speeds.

Lake Loretta is now a focal point for the City of Albany. Flying in the center of the island is an American flag. Various American flags decorate the island on different occasions. On one occasion, the flag given to my family at my father's funeral flew over the island.

SUMMERS AND GRADUATE SCHOOL

I worked on a survey crew every summer from the time I was eleven years old until my graduation from Albany High School in 1956. I was a four-year honor graduate from Albany High School, and, after graduation, I looked forward to studying civil engineering and land surveying at the Georgia Institute of Technology in Atlanta, Georgia. I enrolled in Georgia Tech September of that year.

The civil engineering department at Georgia Tech required all civil engineering graduates to take several courses in surveying and also attend a summer surveying camp. The purpose of the surveying camp was to give students experience in actual field surveying.

The department divided us into three-man crews and gave us one week to complete a survey project. My crew consisted of myself, Edmund Glover, and another student whose name I cannot recall. As luck would have it, both Edmund Glover and I had worked several summers on a survey crew. We were familiar with all of the equipment, and both of us were very efficient in taking notes and performing topographical surveys. We were the only students with experience of that nature.

Just a few days into the project, our crew was so far ahead of the other crews that the professors required us to take a day off without working in order to let the other crews catch up. We all really enjoyed the holiday from school, and when we got back to work, we still completed the project ahead of all the other crews.

Only one person received an "A" for the course, and that was Edmund Glover. He was also recognized as the outstanding student at the survey camp. Edmund and I became close friends after the camp, and, more than fifty years later, we still exchange Christmas cards, although we have not seen each other in many years.

After graduating from Georgia Tech with a bachelor's degree in civil engineering, my parents encouraged me to continue school and get a master's degree. They suggested the degree be in a field different from civil engineering. I decided on city planning.

There was a problem. My undergraduate grades were not high enough to allow me to enter graduate school at Georgia Tech. I had done well my freshman year, but my sophomore year was a disaster. Although I had never made less than an "A" in any physics course and had made an "A" in my previous calculus course, I flunked both physics and calculus the beginning of my sophomore year, due to some temporary health conditions.

Other schools besides Georgia Tech offered a master's degree in city planning. Those schools would allow me to enroll in the program. I don't know why I didn't. I think it

COLLEGE AND MILITARY

was just that I did not like the idea of being turned down by Georgia Tech.

I arranged a meeting with faculty of the Georgia Tech city planning department. The head was Howard K. Menhinick. One of the other professors was Malcolm Little. Both told me I did not have the grades to justify them allowing me to pursue the master's degree program. I was not about to give up, and, after about half an hour of conversation, they finally said, sarcastically, "If you will take five undergraduate courses that we select, and if you make no more than one "B" and an "A" in the remaining four courses, we will allow you to enroll."

They thought that was an easy way to get rid of me, but to their surprise, I agreed. I do not remember the name of all five courses, but Statistics, English, and Psychology were three of them.

I was determined to make it. I rented a private room in the YMCA, which was located on campus. I set up a schedule where I studied a minimum of two hours for every hour in class, sometimes three hours for every hour in class.

Georgia Tech was on the quarter system at the time, with each quarter lasting three months. At the end of the quarter, I had made an "A" in all five courses. The city planning instructors were pleased and allowed me to enroll.

That quarter, I had been allowed to attend school as a provisional student, and after making all "A's," the system placed me on probation. The next quarter I made all "A's" with the exception of one "B." This allowed me to become a regular graduate student. In order to catch up with my class,

I went to the registrar's office with the request to take 21 hours of coursework the coming quarter.

The registrar took a quick look at my folder and told me any student just returning from probation should not be requesting to take 21 hours of coursework. I felt confident in my position and thought I would have a little fun with the registrar.

"I'm really sorry I made that one "B" over the last two quarters," I said. "If I promise to study really hard, will you please let me take 21 hours of coursework?"

"You are just returning from probation and asking to take 21 hours of coursework," he said. His voice was sarcastic, and his expression showed disgust.

"Just look at these grades," he said, slamming my report card down in front of him so that he and I could both see. His right index finger covered the one "B" out of two quarters of coursework. All he saw was a string of "A's." His frown turned to a confused grin as he asked what was going on. I explained that I had been required to prove myself before entering graduate school. Starting as a provisional student, followed by another quarter on probation, was the requirement I had to meet before qualifying as a regular graduate-school student.

I received my master's degree in city planning, graduating number two in my class. My professors, especially Professor Little, enjoyed telling their friends that I was the only student they knew who had to make five "A's" in a single quarter in order to get put on probation.

Some tell me that I set a record at Georgia Tech for the largest spread in grade-point averages during two different quarters. The quarter I flunked physics and calculus, I had a 1.1 grade point average for that quarter, out of a possible 4.0. Later, I had another quarter in which I had a 4.0 grade point average. My dad enjoyed joking about the fact that I flunked more than half of my courses one quarter and made all "A's" another quarter. My mother was always embarrassed that I ever did so badly in school.

Although I graduated from high school in 1956, and although I received my bachelor of civil engineering degree from Georgia Tech in 1960, I did not receive the degree of master of city planning until 1966. That degree required two years of coursework plus a thesis. I completed my two years of coursework in 1962. Two professors approved my thesis at that time. To graduate, I needed approval by three professors.

My thesis was entitled, *A System of Open Spaces for Outdoor Recreation in Metropolitan Areas*. The head of the department, Mr. Howard Menhinick, felt I needed just a few more revisions to my thesis.

I had enrolled in ROTC during my time in Georgia Tech and was required to serve two years with the US Army Corps of Engineers. The government granted me a two-year extension in order to complete my master of city planning degree, but they denied three additional months to complete my thesis. My thesis of more than 100 pages, which needed only the approval of the head of the department in order for me to graduate, had to wait.

I had to complete my obligation with Uncle Sam before completing all graduate-school requirements, but I did not want to wait any longer before marrying my high school sweetheart, Fonda Gayle Starnes. We married on June 16, 1962. Frederick Wilson was the pastor. We married in the First Methodist Church of Albany, Georgia.

Marrying Fonda was one of the best decisions in my life. I have said many times, if I could do it over again, I would have married her many years sooner. I felt that before I married, I needed to complete college, have a good job, and have the income necessary to support a wife. Now I realize none of this is necessary. What is necessary is for the husband and wife to love each other, be faithful to each other, and desire to spend an eternity with each other. As of this writing, we have been married more than fifty-seven years. It's a wonderful life.

SAND DUNES

During my college days, and before I entered the military, I worked each summer with my dad. One project was a topographical survey of the sand dunes in Albany, Georgia. The sand dunes, probably formed by erosion of sand from the Flint River, extended about five miles along the Flint's eastern bank in Albany, Georgia. They were located near the center of town just south of Oglethorpe Boulevard. In my college years, they were more prominent than they are today, but they still exist.

Sand was ten feet deep, sometimes more. Driving an automobile across the sand was hazardous. I have had cars stuck in mud many times, but that is nothing like being stuck in sand. Sometimes the car would sink so deep into the sand that the only way to recover the car would be to get a tractor

COLLEGE AND MILITARY

to pull you out—providing the tractor did not get stuck, also. Dune buggies enjoyed riding across the sand, and, usually, a car with four-wheel drive could navigate the terrain, but the normal car would bog down if it tried to cross.

Walking on the sand was equally hard. This was not like walking on soft sand on a beach. The sand was soft. That was true. It was just too soft and too deep. Sometimes my feet sank more than a foot into the sand, and when I lifted my foot, my shoe remained stuck. The summer temperature while surveying was often more than one hundred degrees. With no shade, it felt like two hundred.

I performed the topo using a plane table and a telescopic alidade, as described in the section on Darton State College. This was a tedious and challenging process. Carrying a drawing table across sand was difficult enough, but trying to keep the drawing dry was another challenge. With temperatures often 100 degrees Fahrenheit or more, I would sweat profusely. Sweat would drip from my face, hands, and arms onto the paper drawing, wetting the paper and making the preparation of an accurate map extremely difficult.

I prepared final drawings in cooler conditions, inside our office. We did have small window air conditioners then, although they were not as efficient or as effective as the central air conditioning we enjoy today. I don't know where the final drawing of the Albany, Georgia, sand dunes are today. I hope someday someone finds them and preserves them. Those drawings would be part of the history of Albany, Georgia, not available from any other source.

MY SON, NO MY SON

During my college years, one of my greatest enjoyments was working with both my father and my grandfather. My grandfather started in the engineering business in 1913, the year my father was born. That year, he called the firm "Marbury and Wright." His partner was another engineer named C.Q. Wright, who later became city engineer for Albany, Georgia. After C.Q. Wright left the firm, the name changed to Marbury Engineering Company in 1935. I was born three years later, in 1938.

The summers during my college years afforded many opportunities to work together as a family. Dad was active in the state surveyors' association and later became president. At times, dad gave lectures at state association meetings. Sometimes I sat in the audience.

My dad, my granddad, and I all had the same name. Dad was Ritchey McGuire Marbury, Jr. Granddad was Ritchey McGuire Marbury, Sr., and I was Ritchey McGuire Marbury, III. Sometimes it got confusing when we received calls asking for "Mr. Marbury." For a short time, we even had company stationery showing the names R. M. Marbury Sr., R. M. Marbury Jr., and R. M. Marbury III.

One afternoon, while I was still in college, a client called to speak to my dad. That was when the fun began. My granddad answered the phone. The client went to great lengths to explain his problem. Granddad said he had no idea how to solve the problem.

"I know you are aware of how to solve this problem. You talked about this issue in your presentation at Georgia Tech just a month or two earlier," said the client.

COLLEGE AND MILITARY

"I made no presentation at Georgia Tech," explained granddad.

"Is your name not R. M. Marbury?"

"Yes."

"Then you made the presentation."

"You must be talking about my son. His name is also R. M. Marbury, except he is R. M. Marbury, Jr."

"No," said the client. "Your son was in the audience."

By now, my dad had picked up another phone in the office and could hear the conversation.

"That was my son," explained dad. "His name is also R. M. Marbury, except he is R. M. Marbury, III."

Everyone had a good laugh. My dad was able to solve the problem, and we all went back to work. Years later, in 1967, I had a son and named him Ritchey McGuire Marbury, IV. For several years, my son and I worked together with my dad, until my dad died in 1994.

JOHN HERNDON LOST HIS GLASSES

One summer morning, John Herndon, a party chief with our company, left the office headed to Catalina Beach Subdivision, a new subdivision in Albany, Georgia. There was nothing unusual about that day. Work started with John and his survey crew measuring distances along the street. He left the office about 8:15 that morning and arrived at the site about 8:30.

John began setting up his transit while Willie Mitchell, our front chainman, began undoing the surveyor's chain and letting it unfold for one hundred feet along the street. For those not familiar with survey procedures, a chain was a metal measuring device one hundred feet long with markings every foot. The first foot of the device had markings every one tenth of a foot. That was before the days of electronic distance meters or GPS equipment.

Sometimes while measuring in streets, drivers became irritated at the delay caused by the surveyors. That day was one of those days. An angry driver sped down the street, pulled out a gun, and fired. John was not able to tell whether the driver used a pistol or a rifle, but he did know someone had shot at him. They missed John, but not his glasses. I don't remember if John was wearing sunglasses or prescription glasses that day, but I do remember that his glasses were shot off by an angry driver.

The interesting thing about that day is that I didn't find out what had happened until the afternoon. John just went about completing the survey and came back to the office after he completed the job. He treated having his glasses shot off as just another day of surveying. I did not have the same reaction. Nevertheless, John and his crew went back the next few days to finish the job. There were no more shootings.

PAULIN ALTIMETER

As I have stated many times, surveying was different in the late 1950s and early 1960s from what it is today, in the twenty-first century. We had no total stations or GPS equipment. If we wanted to know the elevation of some point a mile or two away, we had to start measuring from a known point and measure elevations from the known point to the desired location. Then, to verify the accuracy of our measurements, we

COLLEGE AND MILITARY

had to measure from the desired location back to our original point, or benchmark. Sometimes this could take several days, especially in heavily wooded terrain.

My father found a way to accomplish this task in a matter of hours. He purchased an American Paulin Altimeter. The Paulin Altimeter was similar to altimeters used by pilots, except even more accurate. The Paulin Altimeter purchased by my father could measure differences in elevation to an accuracy of slightly less than one foot.

On one occasion, a client asked us to determine if a sanitary sewer line could be designed to flow by gravity from one point to another point approximately two miles away. We had only one day to find the answer. The sanitary sewer line would be constructed through a heavily wooded area. There was no way to accomplish this task based on conventional surveying methods. Dad decided to use his Paulin Altimeter.

Dad used two altimeters. He calibrated both to precisely the same elevation. He placed one altimeter on the ground at the base point or benchmark. This altimeter never moved during the entire survey. If the reading on the base altimeter never changed, he recorded the elevation on the roving altimeter as noted. If the reading on the base altimeter changed, the operator on the base altimeter called the roving altimeter operator with a handheld walkie-talkie. Adjustments in the roving altimeter were made accordingly.

Dad took elevation readings at various points along the route, taking a final elevation reading at the destination location. Although exact locations were known only at the beginning and ending points, and although we understood that the accuracy was good only to within a little more than one foot, the information was good enough to determine that a gravity sewer could be designed in the location desired.

Today, with modern-day GPS technology, this would be a simple matter, with accuracies far greater than those obtained with the Paulin Altimeter. Nevertheless, the Paulin Altimeter was innovative for its time and solved many needs for quick solutions to problems of elevation differences over great distances.

MILITARY

Since the military would not grant me enough time extension to complete my master's degree, after December 1962 it was time to enter military service. The first week of January 1963, I arrived at Fort Belfour, Virginia for basic training as a second lieutenant with the US Army Corps of Engineers. Fonda went with me, and we found a place to live in Mount Vernon, Virginia. It was a small cabin behind a home owned by a local family.

The Army did not like me bringing my wife to live with me while I was in boot camp. They also did not like my living away from the barracks. Nevertheless, this is what Fonda and I chose to do.

Basic training was not a fun experience. I arrived early in the morning for drill, calisthenics, and training. I felt the training was primarily harassment, but then we were preparing for the rigors of combat, and I knew that anything I experienced here would be much less difficult than any experiences I might deal with in actual combat.

COLLEGE AND MILITARY

While I did well in most of the training, I did not do well with inspections. In fact, Fonda became an expert in spit-shining boots and would shine my boots in the evening so that they would be presentable for inspection the next morning. Fonda was a great help, and my boots always passed inspection. Still, there were inspections that I failed.

One morning in particular, I was running late and found my nametag had come loose on the left side, where the stitching had unraveled. Not having time to sew the nametag back onto my shirt, Fonda found a piece of bubblegum and used that to attach it back to my uniform. It looked fine to me, but during inspection the inspecting officer noticed the loose name tag and ripped it from my shirt.

"What is this?" asked the inspecting officer. "Chewing gum?"

"No, sir," I replied. "Bubble gum."

As expected, I did not pass inspection that day.

During the first two weeks of training, I became sick with a temperature of more than 103 degrees. Because of this, I could not meet all of the physical-fitness requirements. Not only that, during the time I was sick, all trainees were required to visit the infirmary to check their physical condition. I went, acting as if I were in peak condition.

The medics stuck a thermometer in my mouth and placed me in a line where other medics would later remove the thermometer and record my temperature. I knew that if they found out I had a temperature of 103 degrees, they would put me in bed until I recovered. Then I would have to start basic training again at the beginning. That I was not going to do.

Just before I reached the medic, I removed the thermometer from my mouth, shook it until the reading was below 98 degrees, and put it back in my mouth. The medic removed the thermometer, looked at it, said my temperature was just fine, and I proceeded out the door. I felt terrible, but at least I was not going to have to repeat basic training. At least, I believed that at the moment.

I recovered within a few weeks but failed to pass most of my physical-fitness requirements. The lieutenant in charge of basic training, Lieutenant Shuba, reported me to the company commander with the recommendation that I be required to repeat basic training. The company commander then took me outside and ran me through a series of physical-fitness and marching exercises.

By that time, I had recovered and did 14 chin-ups and more than 50 pushups, and I performed every marching exercise with precision. The company commander said my performance was certainly better than reported by Lieutenant Shuba and asked if the lieutenant had anything against me personally. I told him I didn't think so and that the lieutenant probably had me confused with someone else. I was allowed to continue with my basic training.

One incident made me a favorite with my trainers. I say "favorite" sarcastically. We were having bayonet training, and the trainer told us that whatever happened, we must never allow anyone to take our weapon. We practiced various maneuvers, and each time we lunged forward with our bayonets, we were instructed to yell, "*Kill!*"

COLLEGE AND MILITARY

Training went well until my trainer came up from behind me and grabbed my weapon. Without thinking, and with a reflex reaction, I rammed the butt of my rifle into his stomach, flipped him over my head, threw him on the ground, stuck the point of my bayonet at his neck, and yelled, *"Kill!"*

I received special attention from that trainer throughout my basic training. Anything he could do to make my life miserable, he did, but it was worth it.

We also went through escape-and-evasion training. During this training, our platoon was positioned at one point in a field, given a map, and told we must reach a specified location by the end of the day. The idea was that U.S. Army Rangers, who were placed throughout the path from our location to our final destination, would capture us and interrogate us.

The interrogation consisted of being confined to a chair while angry German shepherds growled at us from less than six inches away. They were tied in a way that they couldn't reach us, but we couldn't be sure. My companion and I did not want to go through this ordeal and determined we would not be caught.

I observed the path taken by most of our platoon members and decided to go in a different direction. Our path was several times as long as the path taken by the other members, but we were not caught. In fact, when we reached our final destination, we proceeded to the commissary and purchased ourselves a bottle of soda.

We wandered back into camp later that evening and discovered that we had been reported missing. An all-out search was being conducted to locate us. When they discovered we had been enjoying a bottle of soda at the commissary, our trainers were displeased, but one secretly congratulated us on being the only ones in the platoon who were not captured.

At the end of basic training, my assignment was to remain at Fort Belfour, Virginia, while another lieutenant was assigned to Fort Knox, Kentucky. Fort Knox, Kentucky, was the home of the 538th engineer battalion, and the person assigned there would have an opportunity to work on engineering and surveying projects for the remainder of his tour of duty. The Fort Belfour assignment would consist of training new recruits.

I wanted to do surveying and engineering. The other lieutenant wanted to do training. After a short discussion with our commanders, they switched our assignments. They assigned the other lieutenant to stay at Fort Belfour, and I received orders to report to the 538th engineer battalion at Fort Knox, Kentucky.

I spent the remainder of my military service at Fort Knox, Kentucky. It was a great experience and a boost to my surveying and engineering career. Very near the beginning of my service, I received the assignment to serve as the civil engineering officer as well as the S2 and S3 officer at battalion headquarters. The battalion S2 and S3 officer was the battalion intelligence and operations officer. A major normally served in this position, but since no major was available to serve, I received the assignment. I was promoted to first lieutenant after being on active duty for six months.

My grandfather died on November 7, 1963. He had been sick with lung cancer for many months, and previous to his death, I had requested permission to go home and visit him.

COLLEGE AND MILITARY

Permission, however, was denied. When he died in November, however, I was given permission to return home in order to attend his funeral. Few people have the privilege of spending as much time with their grandfather as I was able to spend with mine. I went fishing with him. I went hunting with him, and we worked together during summer months for more than ten years.

As stated earlier, my grandfather's office prepared the original map of the City of Albany, Georgia, which was officially adopted on June 8, 1926, and revised in June of 1930. He also prepared a map of Dawson, Georgia, sometime during his early years as a land surveyor and engineer, which I updated on September 12, 1967. I researched and compiled the updated map, Gary D. Wisham drew the map, and my father checked the map. I am grateful for the many years I had to spend with both my father and my grandfather.

Shortly after arriving at Fort Knox, the post engineer requested I visit with him. He told me he was aware that I had already passed my Engineer-in-Training exam (this exam is now called "Fundamentals of Engineering") and that I had worked with an engineering company for more than five years. He asked what my military goals were. I explained that my goals were to get as much surveying and engineering experience as I could so that my time in the military would count toward experience in allowing me to take the professional surveying and engineering exams.

The post engineer was a lieutenant colonel and a registered professional engineer. He told me his goal was to be promoted to full colonel as quickly as possible. We made an agreement.

MUD, MANHOLES, AND MACHETES

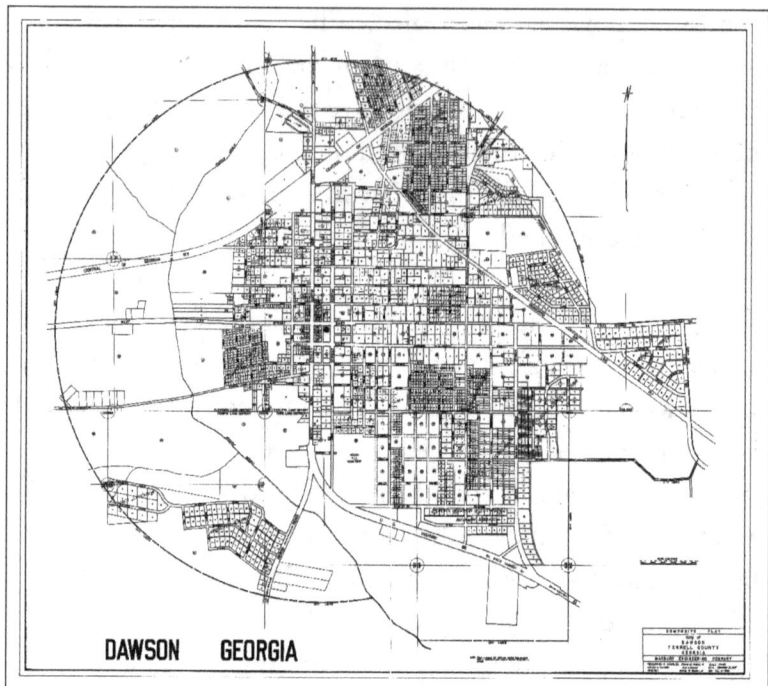

OFFICIAL MAP OF DAWSON, GEORGIA SEPTEMBER 12, 1967 CHECKED BY RITCHEY M. MARBURY, JR. RESEARCHED AND COMPILED BY RITCHEY M. MARBURY, III

COLLEGE AND MILITARY

I could have anyone, including any graduate architect or engineer who was assigned to Fort Knox, to work with me on my staff, provided they were not commissioned officers. I could also select any surveying or engineering projects I felt were needed by the base. In return, I was to give him credit for any successful projects we completed. This was a win-win agreement for both of us and certainly beneficial to Fort Knox.

During my service as civil engineering officer, our team designed bridges, earthen dams, roads, a few buildings, a gas chamber (for training purposes only), and a multitude of other engineering projects. We also performed surveys throughout the base. We were so successful and prolific that our performance came to the attention of officers at the Pentagon. One day a colonel from the Pentagon in Washington, DC, approached me.

"How are you and your team able to produce such outstanding designs in such a short time?" the colonel asked.

"It's because our post engineer is such an outstanding trainer and motivator," I replied. "I don't understand why he is not a full colonel."

"I don't, either," the colonel replied. "I believe I can correct that."

A short time later, the army promoted the post engineer to full colonel. When I received my honorable discharge from the military in January 1965, true to his word, the post commander wrote a letter to the surveying and engineering board of registration in Atlanta, Georgia. He explained the various projects I had designed during my service as the civil engineering officer at Fort Knox, Kentucky.

The letter was in sufficient detail that the registration board allowed full credit for my military service to count toward surveying and engineering experience. I later passed both the engineering and surveying exams on my first try and was one of the few individuals ever allowed to count military service toward engineering and surveying experience.

October 1963 was a special month for me. Fonda and I were expecting our first child, and I was as nervous as a cat in a fire-ant bed. My greatest fear was that the baby would be born before I could get Fonda to the hospital. I was terrified that I might have to deliver the baby myself. Fonda, on the other hand, did not seem to be nervous about anything.

Bowling was a popular sport at Fort Knox, and I was one of the members on the 538th engineer battalion bowling team. One evening during the first week in October, I was participating in a bowling tournament. An announcement came over the loudspeaker just as it was my turn to bowl. The announcement said that Lieutenant Marbury had a phone call. This had to be a call saying my wife was delivering our first baby. I was not there to take her to the hospital. I jumped over three rows of chairs and sprinted to the telephone.

When I answered the phone, the voice at the other end of the line was simply a private asking for a weekend pass. I approved the pass and stumbled back to the bowling alley. When it was my turn to bowl, I rolled the ball right down the middle of the gutter. I was as nervous as a stray dog in a thunderstorm. I think I scored somewhere between ninety and ninety-five that evening and got hardly ten minutes of sleep that night.

COLLEGE AND MILITARY

A few days later, at approximately five o'clock in the morning, my wife calmly said she thought it was time for us to go to the hospital. She said her water had broken and that she was about to have a baby. Five minutes later, I was ready to go. Fonda took another hour or two. She didn't want anyone to see her, even at the hospital, without her hair fixed and her makeup on.

We owned a Plymouth Valiant in those days, with a large citizens-band radio antenna attached to the rear bumper. The antenna stretched from the rear bumper to just above the left front window of our car. A small hook held it in place. When I wanted to use the CB radio, I would unhook the antenna so that it stood straight up from the rear bumper.

A lady down the street told us that, when we went to the hospital, we were to unhook the antenna so that it would stand straight up in the air. This would be a signal that Fonda was having her baby. I unhooked the antenna to signal our friend that the baby was on its way, and Fonda and I sped to Ireland Army hospital, Fort Knox, Kentucky.

At 4:34 p.m. (that was 1634 hours army time) October 11, 1963, our first baby was born. I had no idea if it was a boy or girl, and the hospital would not tell me. Before I was allowed to see my wife, I was allowed to see our baby through a glass window. The baby was in a bassinet with a pink ribbon, and it never occurred to me that the hospital used ribbons to indicate the sex of the child.

When I finally saw my wife and asked her if the baby was a boy or girl, she smiled and told me we had a beautiful baby girl. She also laughed when I told her that, when I saw the pink ribbon on the bassinet, I still did not realize that we had a baby girl. We named her Mary Kathryn Marbury and gave her the nickname "Mitzi."

I remember the first time I held my daughter. I was afraid she would break; I touched her on the tip of her forehead with my index finger. Then I exclaimed, rather nervously, "I touched her."

Fonda stayed in the hospital for several days. The military hospital had strict visiting hours and strict rules of conduct during visits. On one occasion, before the hospital allowed Fonda and our baby to come home, I was visiting Fonda in the hospital. There were no chairs in the room, so I sat on the side of Fonda's bed as we talked together. I was in the room less than five minutes when an army nurse with the rank of major entered the room.

"Get off that bed," she said sternly and continued to reprimand me. "What do you think this is, the French Riviera?"

"This is my wife," I said. "We just had a baby."

"I don't care what you had," the nurse replied. "Get off that bed."

I did.

A few days later, I took Fonda and our daughter home to our apartment at 5650 B Demoret Avenue, Fort Knox, Kentucky.

I was fortunate during my two years in the military. I spent most of my time doing surveying and engineering projects, and never saw combat. There were two times, however, when I could have easily been involved in intense fighting. The first of these times was during the unofficial Cuban missile crisis.

COLLEGE AND MILITARY

I was having a routine day, looking at possible civil engineering projects, when I was ordered to report to battalion headquarters. The government was aware of missiles located throughout Cuba. They determined to send in military to destroy these missiles, and I was scheduled to be among the first to depart. The Marines and U.S. Army Rangers were to arrive first. I was in the second group. Our mission: find the best location to construct an airfield.

The remainder of our construction battalion was to arrive a few days later. By that time, my team was to have determined the best location for an airfield, develop plans for its construction, and be ready to put the remainder of the construction battalion to work within hours of their arrival.

The government issued me several bandoliers of ammunition for my M1 carbine rifle. I knew this was serious the moment I received live ammunition. I packed my battle gear, said goodbye to Fonda, and prepared to board a train bound for southern Florida. From there, my team and I would to be airlifted into Cuba.

Preparation stopped abruptly, and all of us wondered why we were not boarding the train. We learned why a short time later. The Russians had loaded all of the missiles onto ships and were departing Cuba. There would be no attack by U.S. forces. Our mission into Cuba was canceled.

This event never seemed to get much publicity. The official dates for the Cuban missile crisis were October 16, 1962 through October 28, 1962. This was before I joined active military service. The event just described happened near the middle of 1963, just a few months before the assassination of President Kennedy on November 22 of that year.

The second time I came close to being involved in actual combat was during the last two months of my service in the military. The battalion commander called me into battalion headquarters and asked if I would consider extending my tour of duty. I replied that I volunteered for the military because I felt serving my country for two years was the right thing to do. I had no desire to be a regular army officer or to make a career of the military. I wanted to return home as a civilian and begin my career as a professional land surveyor and civil engineer.

The battalion commander told me that our battalion could be going to Alaska soon and that the experience would be good for my career. He smiled, and I smiled. I asked why we were receiving so much mosquito netting, lightweight clothing, and jungle gear.

"We're going to Vietnam, aren't we?" I asked.

He said if I told anyone, he would see that I served at least two additional years. I told him I would be completely truthful with anyone who asked and explain that he told me we could be going to Alaska very soon. The battalion commander said he was glad I understood. We saluted each other, and I left the room.

There was still the possibility that our battalion would go to Vietnam prior to my release. As soon as the news got out, some started finding reasons why they couldn't go. One young private in particular told me he was a conscientious objector and did not believe in carrying a weapon. We went together to discuss this with the battalion commander. At that time,

COLLEGE AND MILITARY

it appeared we would receive orders to go to Vietnam prior to my discharge. This meant that I would have no choice but to remain in the military and serve in Vietnam.

The young private made his case dramatically about why he was a conscientious objector and should not be required to carry a weapon. The battalion commander looked at me and asked for my recommendation. I explained that I wasn't sure I could shoot anyone, either, and, because our main responsibility was to construct an airfield, the young private could be my assistant. Neither of us would carry weapons, and we would be able to focus our total attention on designing and constructing the airfield. After all, I explained to the young private, the Rangers and Marines would be there to protect us.

The young private showed anger immediately. He told us both that, if he went to Vietnam, there was no way he was going to be told that he couldn't carry a weapon. He said if anyone was going to shoot at him, he was going to shoot back. I turned to the battalion commander and told him he had his answer. It seemed our young private was not a conscientious objector at all.

My battalion did not go to Vietnam prior to my discharge. About three months after my discharge, however, it did go. I was ready and willing to go should they leave prior to my discharge. Still, I was glad I didn't have to go. Ironically, the young private went to Vietnam along with the rest of the battalion. After the war, all of my close friends returned home without serious injury. I never heard what happened to the young private. I hope he returned home safely, also.

Chapter 4

DAD AND I ARE PARTNERS

HOME FROM THE MILITARY

On January 8, 1965, I received an honorable discharge from active duty with the U.S. Army Corps of Engineers after serving for two years. Dad made me a partner in Marbury Engineering Company and vice president of the firm.

The first few years after returning from the military were tough. My mother was very ill. My daughter, Mitzi, was only fifteen months old. Like most young couples, we had little money. I had two immediate goals. The first was to get my land surveyor's license, and the other was to complete my master's degree.

I submitted all the paperwork to take my registered land surveyor's exam near the middle of 1965. The Georgia Board of Registration gave me credit for all of my military experience, and I took the exam later that year. I passed the exam on the first try and received my license as a registered land surveyor with the State of Georgia on the fifteenth day of December 1965, registration number 1495.

MASTER OF CITY PLANNING

The other goal was to complete my master's thesis in order to receive a master's degree in city planning from the Georgia Institute of Technology. The task was much more difficult than I anticipated. During my service in the military, someone else completed a thesis on the subject of open spaces, and out of more than one hundred pages of material I had written, all but about the first ten pages had to be rewritten. This was a grueling task. I worked with Marbury Engineering Company two or three days each week. The other days, I worked on my thesis. Without my wife's help, I would never have succeeded.

Each day as I worked on my thesis, I would dictate into a reel-to-reel tape recorder. I dictated until midnight or later each evening and then woke up Fonda. Fonda typed what I'd recorded. When she finished typing at about six in the morning, she got me out of bed.

Then I would get into my car and drive for a little less than four hours, and a little more than 180 miles, to Georgia Tech. There I reviewed what I had written with my advisor and drove back home, arriving between five o'clock and seven o'clock in the evening.

After I arrived home, I made the corrections by dictating until sometime after midnight. Next, I would wake up Fonda so she could type the corrections. Around six or seven the following morning, Fonda would again wake me and hand

DAD AND I ARE PARTNERS

me her typed corrections, and I would drive to Atlanta and repeat the process of the previous day.

Around March 1966, I was close to having my thesis approved. Since Professor Menhinick was the one who had previously required all of the changes, I did what was necessary to receive his approval first. Shortly afterwards, I received approval from Malcolm Little, and the only signature left to be obtained was that of Willard C. Byrd, my third reader. Three faculty members were required to approve my thesis, and I now had approval from two of them.

Professor Byrd was a consultant landscape architect who also taught classes with the city planning department. He notified me that he was ready to approve my thesis but that he was leaving shortly to go out of the country. Graduation was June of that year, and if I didn't meet with him the next day, he wouldn't be available to sign my thesis in time for me to graduate in June. I was afraid that, if I missed June graduation, someone might again write on my same subject, and I would again have to start over. I didn't believe I could do that. Whatever it took, I was going to get Professor Byrd's signature the next day.

Early the next morning, I scrambled into my 1963 Volkswagen Beetle and sped off to Atlanta, Georgia, to meet with Professor Byrd. It was a rainy morning, and the winding two-lane road from Albany, Georgia, to Atlanta, Georgia, was flooded with traffic. As I rounded a curve near Smithfield, Georgia, the left rear wheel of my Volkswagen Beetle suddenly went flat. The car spun out of control and skidded to the left side of the highway.

I believed that, if I applied the brakes, I could get the car under control, but a large eighteen-wheel tractor-trailer was seconds away from a head-on collision with my small Beetle of a car. There was nothing for me to do but allow the car to go over the side of the road, landing upside down in the roadside ditch.

It's funny how your mind works in such a situation. At the time, I had no thought of getting killed or injured in an automobile accident. My only thought was that this was delaying me from getting to Atlanta and getting the necessary signature to complete my thesis and receive my degree.

My seatbelt was fastened. I wasn't hurt, but I struggled to get out of the car. As I struggled, I heard voices from a crowd of people standing outside in the rain, looking at my predicament. Several were saying, "Is he dead? Is he dead?"

When I finally pried open the car door and crawled out on my hands and knees, my only concern was still how to get to Atlanta in time to get Professor Byrd's signature.

I yelled at one of the bystanders.

"Help me pick up this car and turn it right side up. Then I can change the tire and get back on the road. I have to get to Atlanta quickly."

Bewildered, one of the bystanders agreed to help, and we quickly turned the car right side up. I had a spare tire in the trunk and, within a matter of minutes, changed the tire. I told the man "Thank you," jumped back in my car, and resumed my trip to Atlanta.

As I rounded the next bend in the road, in my rearview mirror, I saw most of the crowd still standing outside in the

DAD AND I ARE PARTNERS

rain, probably wondering what had just happened. Adrenaline will do amazing things. I had just picked up a car with the help of one small bystander, turned it over, changed a tire, and drove off—all in a period of less than 15 minutes.

It wasn't until after I had reached Atlanta, obtained Professor Byrd's signature, and driven back home that I realized what I had done. I had almost single-handedly picked up a car. I tried to see if I could do it again, but no matter how hard I tried, I couldn't make the car budge even a few inches.

Then I thought about all those people standing in the rain as I drove off. Some of them are probably still telling their grandchildren about this kid who picked up a car with the help of a stranger, changed the tire, and drove away, all in less than 15 minutes. At that time, the fear of having to write my thesis again was greater even than the fear of being killed in a head-on collision with an eighteen-wheel semi.

On June 11, 1966, I walked down the aisle at the Georgia Institute of Technology to receive my master's degree in city planning. My wife also received a degree at that ceremony. It was a degree conferred upon wives of Georgia Tech students who worked so hard—and sacrificed so much—so that their husbands could complete their degrees. Her degree was entitled, "Mistress of Patience in Husband Engineering." Her diploma read,

"Georgia Institute of Technology. Atlanta, Georgia. This certifies that Fonda Starnes Marbury has continued faithfully to support and encourage a husband through many months of general trials and tribulations, including endless conversations

87

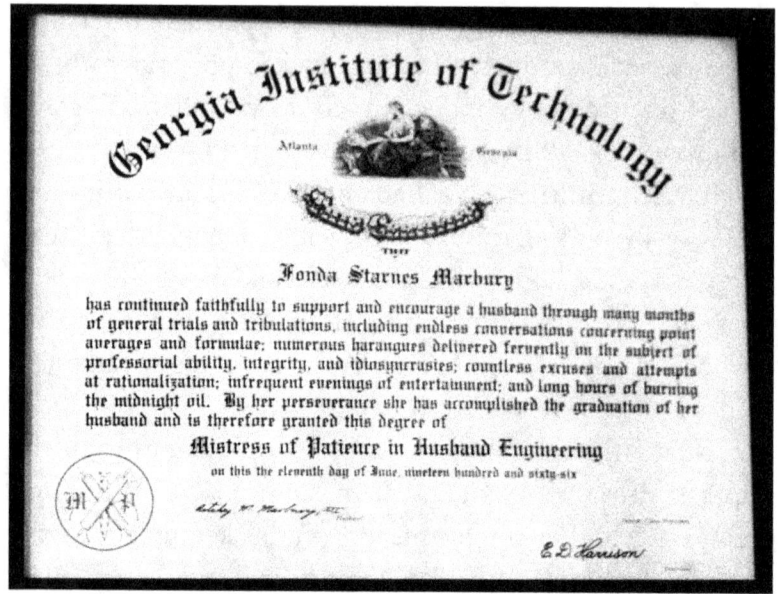

concerning point averages and formulas; numerous harangues delivered fervently on the subject of professional ability, integrity, and idiosyncrasies; countless excuses and attempts at rationalization; infrequent evenings of entertainment; and long hours of burning the midnight oil. By her perseverance she has accomplished the graduation of her husband and is therefore granted this degree of Mistress of Patience in Husband Engineering on this, the eleventh day of June, nineteen hundred and sixty-six."

The diploma was signed by me, Ritchey M. Marbury, III, as husband, and by E. D. Harrison, president of the Georgia Institute of Technology. The seal at the bottom-left corner of the diploma consisted of a circle, within which are two crossed rolling pins. On the left side of the rolling pins is the initial

"M," and on the right side is the initial "P." The "M" and the "P" stand for Mistress of Patience.

In the process of spending numerous nights typing drafts of my thesis, Fonda typed more than three thousand pages—and this was on a typewriter, not a computer. We did not have computers with word processing then as we have today, and few people had any knowledge of computers in those days. She did all of her typing on a small portable typewriter.

My college and military days were now behind me. My career working full time as a land surveyor and civil engineer was just starting.

The year 1966 had its good times and bad. The good times, of course, were the fun times working with my dad and the fact that I'd finally completed my master's thesis and had graduated.

The bad times were those late nights and weekends working on my thesis and the fact that my mother was so ill. My mother died on September 25, 1966, at Phoebe Putney Hospital in Albany, Georgia. My dad sent me home that night while he stayed beside her bed, holding her hand for her last time on this Earth.

IF YOU HAVE NO HILL, MAKE A HILL

Life went on, and I wanted to live in a subdivision I had personally designed. My wife, Fonda, had always wanted a house on a hill. She'd spent much of her life, before she married me, in the hills of Tennessee. The countryside overflowed with hills, valleys, and scenic vistas. I would sometimes visit her at her parents' home in Estill Springs, Tennessee. They lived on a hill where one could look down and see rivers, small farms, and railroad tracks. I lived in Albany, Georgia. It's a beautiful place, especially in the springtime, but it's as flat as a squashed paper bag.

During the summer of 1967, Fonda and I decided to build a home in Lake Park Subdivision. This was a subdivision I had designed and was under construction. We selected a lot along the curved part of Green Valley Lane. The front width was about 115 feet, but since it was on a curve, the back was about 300 feet wide. The owners priced the lot based on the front feet, so we paid for 115 feet of frontage but had a lot with more than three hundred feet in the rear, a great value. I designed the street so that it discouraged through traffic—just what Fonda wanted, except for one thing: the lot was flat. Fonda wanted to live on a hill, and the lot was flat.

"Fonda," I explained, "we live in Albany, Georgia, and there are just not many hills here. This is the perfect lot except for the hill. Won't that be OK?"

"I want to live on a hill," said Fonda.

"There is no hill here."

"Then lower the street."

I thought about that for a moment. The lots at the far end of the street were too low to build on, and perhaps I could use the dirt from lowering the street to fill in the remaining lots. I would have to act fast. Street construction was underway.

I recalculated street grades, lowering the street in front of my lot by a little more than four feet. I surveyed and reset the grade stakes for the street and advised the contractor.

DAD AND I ARE PARTNERS

Later, I revised the plans to reflect the new street elevations. I instructed the contractor to use the excess dirt from the street to fill the low lots at the far end of the street. It was done. Fonda had her lot on a hill.

It's interesting how things work out. The public loved the low streets and high lots. Lots sold quickly and at much higher prices than originally planned. The once-unusable lots at the lower end of the street were now prime real estate. My wife was happy. My clients were happy, and I was happy.

SURVEYING IN A COW PASTURE

Every survey site had its challenges. Never was this truer than when surveying in a cow pasture. One spring day, we had a survey project in a cow pasture about ten miles from our office. There were two major problems. One was that the pasture had one bull and several cows grazing right where we needed to survey. I had a healthy respect for bulls and was certainly not a bullfighter. When I looked at the bull, and he looked back at me, I knew I didn't want to be in the same field with him. The owner agreed to remove the bull from the field while I worked on the survey.

The second problem was that the cows were a little too friendly. They were curious about what was going on. I would set up the transit to start the survey, and the cows would move in to look. The first time I set up the transit and was about to measure an angle, a cow bumped me in the rear. I bumped into the transit. I didn't know what had hit me, and I grabbed for the seat of my pants. I got the cow's nose instead. The cow jumped away, and I grabbed for the falling transit. I did manage to catch the transit before it fell to the ground, but I had to set it up again.

This time I watched for cows behind me, but another cow in front of me wanted to smell the transit. She bumped the transit, causing me to have to set it up again. We had a three-man crew. One member was on the back-site point. The second member was on the front-site point. This meant no one was left to help keep the cows away from the transit. Finally, I managed to set up the transit without being bumped by cows and measured the angle.

Surveyors always have to watch out for something wherever they work. They have to watch for cars in cities, snakes and alligators in swamps, and cows in farm pastures.

ALBANY MALL

One summer afternoon, Aaron Aronov came to the office asking our opinion on the feasibility of developing a shopping center on the northeast side of Dawson Road, US Highway 82. He talked to several engineers with the understanding that one of us would be selected to do the site design.

Dad looked at the site and told Mr. Aronov not to develop there. He said that the soil conditions were poor and that development would be very expensive. He also told him that storm-drainage design would be difficult, since no location for the outfall storm drainage existed. He said the stormwater runoff would have to be carried either across Dawson Road and through several subdivisions southwest of the site or northeast, into Lee County. Neither of these solutions were good ones. Again, my father told him not to develop on that site.

DAD AND I ARE PARTNERS

Several weeks went by, and Mr. Aronov again came to visit us. He told my dad he had decided to purchase the site for the shopping center and that our firm had been selected to do the engineering. Dad said, "Why are you selecting us? I told you that site had a multitude of problems."

"If you know what the problems are," said Mr. Aronov, "I presume you know how to fix them."

"With enough money," said dad, "I can fix anything."

"That's the site where I'm building the shopping center," said Mr. Aronov. "You're my engineer. Fix it."

Our firm, Marbury Engineering Company, became the civil engineers and surveyors for the project with Tiller, Butner, Rosa, and Seay as the architectural firm. Sam Butner acted as the lead architect for the project.

Our company knew engineering, but Aronov Realty, the owners of the new shopping center, knew marketing. They knew that this location was the best in the city and that the potential profits from a shopping center in this location far outweighed the high construction cost. They were right.

On October 19, 1972, the Dougherty County Planning Commission voted on the request to construct a regional indoor shopping center on one hundred acres of land in northwest Albany. The planning director and his staff recommended denial. The commission, however, voted unanimously to approve it.

We went to work. We performed boundary surveys, topographical surveys, and subcontracted for geotechnical and soil surveys of the property. Dad asked me to do the storm-drainage design. I studied the project carefully, did pages and

93

pages of detailed storm-drainage calculations, and presented my solution. Dad said my solution would work well, but he thought he would see what he could do, also.

A few days later, my father had a solution that worked better than mine, at almost one half the construction cost. I was amazed at my father's ability to look at an engineering problem and visualize the best solution in such a short time. Not only that, his solution was so simple that I wondered why I hadn't thought of it in the first place. I have since learned that the best solutions usually appear simple but are often the hardest to find.

I mentioned earlier that the site had serious drainage problems. In fact, the final solution was to send all of the stormwater runoff into the adjoining Lee County. Aronov Realty arranged to purchase large acreage from Ledo Properties and construct a large retention pond on the site. All drainage from the mall property went into that retention pond. As my father said, the project was very expensive, but, as Mr. Aronov said, we knew what the problem was and how to solve it. The solution just cost a lot of money.

Near the end of the design process, John Biesel, an Aronov associate, visited our office frequently. One morning, he questioned dad regarding when dad would finish. Dad said he would be through by noon the next day. Early the next day, John was in our office and stood over dad most of the morning while dad worked. At noon, dad placed his drawing equipment on his drafting table and walked to the door.

"Where are you going?" asked John.

"I'm going to play golf," replied my father.

DAD AND I ARE PARTNERS

"But you aren't finished." John said. "You said you would be finished by noon today, and you aren't."

"I didn't say I would be finished. I said I would be through by noon today. If you had left me alone, I would be finished. Since you insisted on bothering me all morning, I was not able to complete the work by noon. Now, I'm going to play golf. I will work on your project again tomorrow, and if you leave me alone, I will finish the work. If you bother me all tomorrow morning, I will again be through by noon and leave to play golf."

Dad then walked out the door and drove to Doublegate Country Club, where he spent the remainder of the day playing golf. John and I looked at each other, neither of us knowing what to do. I felt we would be fired on the spot, but we weren't. The next day John did not show up at the office, and, as promised, dad finished the project.

My grandmother, Dessie Marbury, died on August 14, 1973. We were still working on the site design of the Albany Mall. All my life, I'd called her by the name "Granna." She was a small, cute, trim, and spry little lady who was a wonderful wife, grandmother, and pianist. One time, she gave me a piano-duet book, and after learning a few songs in the book, we would play duets together.

I remember one evening she came to our house, and we asked everyone to demonstrate some talent. Although Granna could play the piano, she didn't want to play for us at that time.

My son, Rick, said, "That's okay, Granna. You can just spell the word "ostrich" as your talent."

"O S T R I C H," said Granna.

"Great job," said Rick.

We completed the site design of the Albany Mall in early 1974. Several site-construction companies bid on the project. Oxford Construction Company was the low bidder and was awarded the contract—but not at first. Another company was the apparent low bidder.

As dad checked the bid tabulations, he found an error in the apparent low bidder's submission. The owners rejected their bid and awarded the contract to Oxford Construction Company. Groundbreaking was in August 1974.

The mall construction was an exciting time. In an effort to ensure quality construction, both city inspectors, Albany Mall inspectors, and Marbury Engineering Company inspectors reviewed construction almost daily. Gary Flannigan was the sanitary-sewer contractor.

The city inspector had previously instructed Gary not to cover any sewer pipe until he'd approved it. One day, after waiting almost four hours for the city inspector to arrive, Gary felt the cost to wait was more than he could afford. He therefore requested an inspection by the mall inspector and, after confirmation, proceeded to cover the pipe.

When the city inspector arrived a few hours later, he insisted Gary uncover the pipe so that he could personally inspect it. Gary said if he did, the city would be required to pay for the extra expense. The city inspector said, "No," and continued to insist that Gary uncover the pipe. The contractor called me to the site to resolve the dispute. Both sides were deadlocked. No solution seemed possible. Tempers flared until I told both parties that I had a solution.

DAD AND I ARE PARTNERS

By now, the loud shouting and yelling by Gary and the city inspector had attracted a crowd of ten or fifteen people. I told the city inspector and Gary that there was an easy way to solve the problem. Gary and the city inspector could fight it out, and we would charge admission to the fight. I stated that we could earn enough money from that event to pay for the cost of uncovering the pipe for another inspection.

The subtle humor calmed the situation for a few moments. Then I reminded the city inspector that the mall inspector was a registered professional engineer and had verified that Gary had installed the pipe correctly. I also reminded the city inspector that he was not a registered engineer and that, if we found the pipe installed correctly after uncovering it, the city could be held liable for the extra expense. After a little more discussion, the argument ended. Gary proceeded with construction of the sewer line, and there was no more talk of uncovering the previously installed sewer line.

On another occasion, we had a lady drafter working with us during the construction. Her name was Neila Cohen. She was a very attractive girl and as talented as she was attractive. One day, she asked me if she could do some outside work rather than just stay in the office. I gave her the assignment to do construction inspections at the Albany Mall. She loved the assignment and went immediately to do the inspections.

She noted several deficiencies in the construction and brought them to the attention of the superintendent. He made some wisecrack that he later regretted about a female inspector. Neila never told me exactly what he said, but she did tell me her reply. She said she simply smiled at the superintendent, told him he could continue to do whatever he wanted to do, but all payment for his work would stop until he corrected the situation. She also added that the correction must be to her personal satisfaction.

When the contractor failed to receive payment for his work, he realized she was not bluffing. He corrected the deficiencies. Neila approved the corrections, and we never had that problem again. From that point forward, whatever Neila told the superintendent to do, his only reply was, "Yes, ma'am."

The Albany Mall opened at 10:00 a.m. Wednesday, August 4, 1976. An estimated four hundred spectators attended the opening ceremony, and the U.S. Marines' Drum and Bugle Corps performed with patriotic selections commemorating America's Bicentennial, including "God Bless America."

The original shopping center had seventy-six stores. As part of the original marketing, the owners took an idea from the popular musical *The Music Man*. Those marketing the mall developed a theme song called "Seventy-six Great Stores in the Albany Mall," sung to the tune of "Seventy-six Trombones." The three major anchors were Sears, Belk, and Gayfers. Construction cost exceeded twenty million dollars.

DOUBLE YOUR BILL

Throughout my career as a land surveyor and civil engineer, most clients were outstanding. In fact, some of my clients became my closest friends. My dad was the same way.

Perhaps my dad's closest friend was also our biggest client. His name was Spencer Walden. We did all his surveying

DAD AND I ARE PARTNERS

and engineering work from the time I was a small lad until Spencer's death. Spencer was one of the largest land developers in southwest Georgia.

One particular project took far more time than budgeted. The survey went through much underbrush that had to be cut and trimmed away. The engineering calculations proved to be much more difficult than we first expected, and the approval process was like a race between a snail and a turtle. Nothing went right.

At the end of the project, dad prepared the bill. He studied it and felt it was too much for the benefit our client had received, so he reduced it. He looked at it again and reduced it again. Finally, he delivered it to our client.

Spencer Walden took one look at the bill and said he would not pay.

"But we spent much more time on this project than we billed you," said dad.

"Tear it up. I will not pay it," replied Spencer.

Dad looked up at his friend and asked what Spencer felt was a fair price. Whatever it was, dad said he would revise the bill to meet that price.

"Tear up that bill. Revise it, and double it," said Spencer. "I know how hard you worked on this project, and I know you have always treated me fairly. I plan always to treat you fairly, also. What you did is worth double what you invoiced, so double your bill, and I will write you a check right now."

Dad doubled the bill. Spencer wrote the check, and we all went away smiling. That is the kind of clients I was fortunate enough to work with for most of my professional career. Most engineers in private practice do not have such good fortune. I consider myself blessed.

WORKING FOR A CROOK

Most clients and contractors I worked with were honest and ethical, but not all. One contractor in particular was a crook and never minded saying so. My dad and I had designed a project in Camilla, Georgia, and this contractor was awarded the contract.

The contractor completed construction of a sanitary-sewer line and asked for final inspection. I drove to Camilla, Georgia, to inspect his work. It was a small project on the outskirts of town, and I was anxious to complete the inspection and return home. That did not happen.

Sewer inspections in those days consisted of what we called "mirroring the line." We did those inspections only on sunny days. That day was sunny.

I took a large hand mirror and climbed to the bottom of the sanitary manhole. The second member of my crew held another mirror at the top.

I inspected the line by having the person at the top reflect sunlight on my mirror at the bottom. Then I reflected the sunlight from his mirror so that I could see down the open sewer line to the next manhole. If I saw a round circle at the end of the line, I knew the sewer line was straight. If the circle was elliptical or some other shape that was not round, I knew the line was not straight. That meant the contractor had to remove the line and construct it properly.

DAD AND I ARE PARTNERS

That day, not only did I not see a perfectly round circle, I saw no circle at all. Since this was an eight-inch sanitary sewer, that meant the contractor constructed the line so crooked that there was a bend in the line of more than eight inches. I climbed out of the manhole and instructed the contractor to remove the line and construct it correctly.

The next thing the contractor did was to offer me money if I would approve the construction. When I rejected that offer, he told me he would reconstruct the line, but he requested that I not inspect the line again after he'd done the work. He requested that I send another inspector rather than do the inspection myself. I replied that not only would I personally do the inspections, I would be there every day to see that this project was done to perfection.

After several months and several failed inspections, the contractor completed the project. I approved his final invoice, and he received payment in a few weeks.

A month or two later, he came to my office. He looked me straight in the eye and said if I saw him bidding on any other job where I was the engineer, I should let him know, and he would remove his bid from the project. I told him that would be a good thing for both of us.

His next comment came as a complete surprise. He said he had several site-development projects, and he wanted to hire me to be the surveyor and engineer on all of his projects. I questioned why.

He said that he was a crook and that everyone knew it. He also said that I was the only surveyor and engineer who had been able to catch him cheating on the job. He said that if he could not get by with cheating, then no one else could. He said he wanted me to treat all contractors working for him the same way I treated him as a contractor.

I told him I was in business to serve the public and would be willing to work for him on one condition only: he would pay for all services in advance. Since he and I both knew he was a crook, getting paid in advance was the only way I would work for him.

He said he'd expected that. In fact, he said that, if I did not require payment in advance, he would have less confidence in my intelligence. I worked for him for several years; he always paid in advance, and we did many projects together.

WATER TANKS

Dad and I worked on the design of many elevated water-storage tanks. Two in particular are worth noting. One was in Americus, Georgia. The other was on Gillionville Road in Albany, Georgia. The work was dangerous back then, as we usually climbed the tanks without any safety devices.

The Americus, Georgia, water tank was an old tank my dad was inspecting. After the city had drained all the water from the tank, dad climbed up a ladder on the inside of the tank in order to inspect it. Another young engineer climbed up behind him.

About halfway up the tank, one of the rungs broke on the ladder where the young engineer was standing. He fell. The fall killed him. Dad managed to hold on to the vertical bars and suffered no injuries.

DAD AND I ARE PARTNERS

A few years later, I climbed up the middle of a newly constructed water tank to make a final inspection. The tank was located on Gillionville Road in Albany, Georgia, and was about one hundred fifty feet high.

I inspected the inside and then needed to inspect the outside. The tank had a door near the top leading from the inside to the outside. I climbed out that door onto a ladder fixed to the outside of the tank. The ladder was attached in such a way that it was movable, and I could hold onto the movable ladder and work my way around the tank. This was the way I inspected the outside.

I worked my way around the outside of the tank, made my inspections, and then moved the ladder back to the opening at the top. Before I could climb back into the tank, strong winds blew me back around the tank. I could not get in the door.

I carefully inched my way back to the door, and again winds blew me around the tank. At that point, I remembered the young engineer who had been killed climbing the water tank in Americus. This was not a good feeling. The third time, I did reach the door at the top of the tank and managed to climb in. From there, I climbed back down the ladder inside the tank to safety.

I think about that experience today and wonder why I was so foolish as to do such a thing without any harness or other safety equipment. Youth is a wonderful thing, but too often we act foolishly and fail to realize the real danger in the way we act. That was the last water tank I ever climbed.

WATER TANK ON
GILLIONVILLE ROAD
IN
ALBANY, GEORGIA

DAD AND THE FLYING INSTRUCTOR

My dad and I both enjoyed flying together, especially on surveying and engineering projects out of town. Dad purchased a Cessna 177, which was a Cessna Cardinal, and we both obtained pilot's licenses. Dad flew for many years as a young man and later paid for my flying lessons. Pilots must certify annually, and, at age 64, my dad went to Ayers Air Service in Albany, Georgia, to get his annual certification.

That day, I left Albany with the company airplane to fly to an out-of-town project. Dad said he didn't mind and was confident about certifying in one of the planes owned by Ayers Air Service. The instructor was a young pilot, mildly arrogant, and one who definitely did not want to certify pilots in their sixties. In fact, he called dad an "old man" and said dad was too old to be piloting any airplane. Dad just smiled and said he was ready to show what he could do.

The instructor looked over the planes available and noticed a Citabria. A Citabria is an aerobatic airplane—what is sometimes called a "tail-dragger," because of the wheel at the tail of the aircraft. Its name spelled backwards is "airbatic," which indicates it is designed for aerobatic maneuvers. This airplane, the young instructor thought, would present sufficient challenge to show how old men should not pilot airplanes.

"We'll take the Citabria," the instructor told my dad.

"OK with me," said dad.

Minutes later, dad and the instructor were in the air. Dad flew cautiously, staying straight and level as he flew over the southern edge of the city and approached a less-populated area. The instructor had dad do a few steep banking turns and one or two stalls. Dad executed each maneuver flawlessly.

"I want to see what you can really do," said the instructor. "I want to see you ring it out." To "ring out" an airplane means to have it perform many aerobatic maneuvers.

"Do you really want me to do that?" replied dad.

"Yes," said the instructor.

Dad smiled to himself. What this instructor did not know was that dad was a former stunt pilot and very familiar with this airplane and its ability to do aerobatic maneuvers. His first maneuver was a barrel roll, followed by a stall, a tailspin, an upside-down flight over the runway, another loop, and then a landing.

When the two departed the plane, the instructor walked around a few seconds, dazed and obviously shaken. All the blood seemed to have drained from his face, and it looked as though someone had painted it white. When he regained his composure, he spoke to dad in a soft voice.

"I have just four words to say: You pass. I quit."

LANDING BACKWARDS

One day, I decided to take our company plane and fly with Fonda to see her parents in Estill Springs, Tennessee. We planned to land in Tullahoma, Tennessee, where her parents would pick us up and drive us to their home. Although the sky was clear, the wind was fierce. I had no concerns, since we had good visibility.

On our approach to Tullahoma, I checked the wind direction and speed. The wind was blowing directly down the runway but was near forty miles per hour. Our plane was a Cessna Cardinal, equipped with an STOL kit. STOL is an acronym for "short takeoff and landing." That meant that the stall speed was very low and that the plane could land at very slow speeds.

I approached the runway and pulled back on the throttle. The plane seemed to hover over the runway, barely moving. I moved the throttle forward to gain more speed. The plane inched its way over the runway, but I could not get it to descend. I pushed more on the throttle and flew the plane until it was over the middle of the runway. Then I pulled back on the throttle, again slowing the airspeed.

Gradually, the plane settled down on the runway. When it touched down, it was actually moving backwards. The headwind over the runway was greater than the stall speed of my airplane. Although my airspeed was thirty-seven miles per hour, the wind speed was forty miles per hour. I landed going backwards at a land speed of approximately three miles per hour.

I probably shouldn't have landed during such strong winds, but Fonda's parents were waiting for us, and since there was a direct headwind and not a crosswind, I felt safe. I did have trouble securing my aircraft before leaving the airport.

The Tullahoma Airport had no control tower but did have a fixed-base operator. When Fonda and I left the plane, he looked at me, shook his head, and said,

"I've never before seen an airplane land going backwards."

"Neither have I," I replied.

MUD, MANHOLES, AND MACHETES

NIGHT TAKEOFF AND LOST POWER

I am forever grateful to my flight instructor, Gary Cooper (not the movie star). His training probably saved my life. Gary taught me always to be prepared for any emergency and especially for power failure. On every training flight, he would pull the power at some point so that he could teach me how to land the plane without power. Sometimes he would even pull the power on takeoff.

My dad and I had a project in Kissimmee, Florida. We flew back and forth to the project several times a month. On one occasion, our personal airplane was not available, so I rented an airplane. I was familiar with the rented plane and so had no concerns about flying it. I flew to the site, completed my work, and prepared to fly back. It was late at night, and another friend asked if I would fly him to Tallahassee on my trip back. I agreed, and we took off.

On the way back, my friend said he usually flew with another pilot, but that pilot had been killed when his plane lost power on takeoff from the Tallahassee airport. I expressed my condolences over the loss of his pilot, and we continued to Tallahassee. After landing, he departed the aircraft, and I prepared to return home.

The Tallahassee airport has a long runway, and I usually taxied to the middle of the runway to take off. It was something generally approved for my small airplane when there was no other traffic at the airport. This time, I thought about my conversation with my friend. I wondered what his pilot could have done to avoid losing his life.

DAD AND I ARE PARTNERS

I decided to practice what I would do. First, I decided to taxi to the far end of the runway. This would allow the maximum runway length for takeoff. I wanted to practice what might happen if I had power failure.

Then it happened. I was in the air and just above the far end of the runway. The airplane lost power. I could see nothing in front of me or to either side. I knew I was in trouble.

It's funny how things go through your mind when you are in an emergency. I did not realize my danger at the time and just felt I was going through another training exercise with my flight instructor.

The first thing I did was to put the plane into a maximum glide slope. That would keep the plane airborne for as long as possible, giving me the maximum time to land. Next, I banked the plane to the left in an effort to execute a one-hundred-eighty-degree turn and land the plane. Then I radioed the control tower.

"Mayday, Mayday. I have lost power and will be on the ground in about ninety seconds. Please clear the runway."

The tower operator proceeded to give me the weather report. I don't know why. I had been flying for hours and certainly knew what the weather was.

"I have lost power and will now be on the ground in about sixty seconds," I radioed. "Please clear the runway."

By then, the tower operator seemed to understand the situation and lit all runway lights. Fortunately, there was a cross runway at about ninety degrees to the runway from which I had taken off. For the first time, I could see runway lights.

"Cleared to land on any runway," came the voice of the tower operator.

"I hope I'll be landing on a runway," I radioed back. "I'll be on the ground somewhere in about thirty more seconds. I hope it's on a runway."

Thirty seconds later, I landed safely on the runway. I would never have made it if I'd had to make a one-hundred-eighty-degree turn, but I was able to land on the cross runway. When the plane came to a stop, I tried again to start the engine, and it started. I was taxiing toward the hangars when the tower operator called again.

"Since you were able to start the plane, do you wish to take off again and fly home?"

"No," I replied. "This plane can sit here forever, as far as I am concerned. I am renting a car and driving home."

I called a car-rental agency and rented a car. I called my wife to tell her my situation and that I would be late getting home. Then, I drove home.

COLLECTIONS ARE A STINKY BUSINESS

When conventional methods of collection do not work, there are often alternative and even more effective methods. Such was the case with an apartment complex where I performed the survey and engineering work. Work was complete and all billing submitted. Shortly after submitting my invoice, the owner came to my office with two checks. One was for the contractor, and the other was for me.

He told me I could give the contractor his check before or after cashing mine. He said there was enough money in the bank to cover only one of the checks. I could decide which one would get their money. He then left the office grinning like a Cheshire cat waiting for dinner. I gave the contractor his check to cash first, and, sure enough, my check bounced—but that's not the end of the story.

A sanitary sewer ran from this apartment complex through another complex before reaching the public street. I previously had told the owner of the apartment complex to purchase an easement from the adjacent property, but he failed to do so. The final plat showed a proposed easement, but the easement was never recorded. The adjacent owner was also my good friend. I will not mention his name.

When all the apartments were occupied, I called my friend.

"Friend," I asked him. "Are you aware that there is a sanitary sewer running through your property from the apartment complex to the north and that all the bills have not been paid for survey work on that property? If you allow that to continue, it could possibly result in a lien on your property."

"What's the matter?" my friend asked. "Didn't he pay you?"

"No, he not only didn't pay me, but he bragged about getting away with it."

"What would you like me to do?"

"Well," I responded, "would you give me permission to block the sanitary sewer at your property line? The official reason will be that you do not want to risk a lien against your property."

"Do anything you want. I'll back you up. You can even double your invoice if you like."

"No," I replied, "I just want to get paid what I am owed."

"Then go ahead," said my friend.

That was the backup I needed. The next thing I did was to call Gary Flannigan, the contractor I helped get paid. I asked him how much he would charge to block the sanitary-sewer line. Gary said he never liked that owner anyway and would pay me if I would allow him to do it. I thanked him and told him I didn't want any pay; I just wanted him to block the sewer line unless I told him to stop. This was a Friday afternoon. I scheduled the work for the following Friday.

With those arrangements made, I called the owner's attorney, Norman Spence.

"My unpaid surveying and engineering bill is a possible lien against my friend's property, since the sanitary sewer line from this apartment complex runs through it. For that reason, the adjacent property owner has authorized me to have the sanitary-sewer line blocked. The contractor is scheduled to start the necessary construction one week from today."

Norman immediately called the owner. The owner said he would come down in a couple of weeks to talk about this situation.

"Fine," I replied. "Just tell all your renters not to flush their toilets after next Friday."

Norman and the owner called Gary, who verified he was going to plug the sanitary-sewer line this coming Friday.

Frustrated, the owner had Norman call me to say that, if I would give him a letter stating he was paying me as a gesture of goodwill and not because he was being forced, he would pay me. I agreed, and he paid me one day before the sewer was scheduled to be plugged.

When Norman delivered the check, he laughed and told me I was the only one paid through goodwill efforts. He said many who were involved in that project never got their money.

NOT-SO-ORIGINAL MARKER

Surveyors generally always strive to do accurate work, and part of that work involves finding original survey markers. One day a client asked me to do a survey in Camilla, Georgia, on a parcel of land with disputed boundaries. The legal description was clear, providing one could find the starting point.

The legal description described the starting point as an iron reinforcing rod located at the centerline intersection of two local streets. The streets were paved, but I believed I could find the reinforcing rod using the iron finder I always kept in our survey vehicle. The iron finder was called a "dip needle." It was actually a compass mounted vertically inside a small glass encasement. When placed over a metal object, the needle would point toward the ground in the direction of the metal object. I found no iron rod.

I measured from many directions, estimating the street centerline from existing pavement. I found iron pins at lot corners and used those corners to help find the original corner. It just was not there.

Next, I attempted to survey the property based on existing lot corners. I found many up and down the street. The problem was that I found different solutions based on where I started and which iron I used as a starting point. The measurements between existing iron pins did not match the measurements shown on the recorded plats. My client hired me to resolve the disputed property line, and I couldn't find a starting point that I could prove correct.

Next, I began to research my own company's records. Both my grandfather and my father were surveyors before me, and I had all the records. My grandfather was no longer living, so he was not available to help. My father, however, was still active in our company and offered to help with the research. Finally, my dad found notes from when my grandfather had surveyed the same property many years ago. While the starting point, the iron reinforcing rod referred to in the deed, was shown to be several hundred feet *north* of the property, my grandfather's notes referenced several monuments located several hundred feet *south* of the property.

I found the monuments shown on my grandfather's field notes. As I measured up and down the street, all monuments south of the property fit both my grandfather's notes and the recorded plats. Finally, I was sure I had the correct location of the disputed boundary.

In order to help future surveyors, I continued my survey to the location of the original monument referenced in the deed. Still not finding the marker, I drove another iron reinforcing rod into the pavement at the correct location of the original monument. I drove the iron about one half inch below the pavement so it would have less chance of being disturbed.

When I showed my client where I believed the disputed line was located, he agreed. His neighbor did not. His neighbor insisted on hiring another surveyor in order to get a second opinion. Within one week of starting his survey, the other surveyor stated he had found the original iron. He suggested he take his crew, I bring my crew, and together we measure to the disputed line. Both parties agreed to accept wherever our crews placed the corner.

I was surprised that someone else was able to find the original iron but glad that we now had the conclusive proof needed to verify the location of the disputed boundary. When we arrived at the site, the other surveyor pointed with pride to what he emphatically stated was the original iron marker. He also explained to me that the reason I had not found the marker was that it was located below the paved surface. You guessed it. The marker he found was the iron reinforcing rod I'd placed in the road centerline the week before. He even pointed out that the iron was rusted, proving it was the original iron.

Who was I to argue with another surveyor who had conclusive proof that he had found the original iron marker? We proceeded to measure from the found iron to the disputed property line. With confidence, the surveyor announced that my survey was correct, and all parties accepted the location. Almost twenty years later, I confessed to the other surveyor what I had done.

MILLER APARTMENTS

Never judge a potential client by appearances. Larry Walden walked into my office one morning bringing with him a raggedly dressed man in overalls and a ragged shirt. He was pleasant but displayed a seemingly uncultured manner. The man was Charles Miller.

Charles said he wanted to construct an apartment complex near Nottingham Way and Stuart Avenue. The project needed a topographical survey and site design. It also needed rezoning. I looked over at Charles and concluded he could never pay our surveying and engineering bill—much less afford to construct an apartment complex. Nevertheless, due to my friendship with Larry Walden and our longtime relationship with his real estate company, Walden and Kirkland Realtors, I agreed to take the job.

We completed our portion of the project, and Charles was ready to start construction. I asked Charles who his contractor would be.

"I plan to do the construction myself," Charles replied.

"From where will you get the money?" I asked.

"Oh," replied Charles, "I'll be my own banker."

"What about insurance?" I asked. "Who will be your insurer?"

"I have my own insurance company," Charles replied.

The project proceeded on schedule, and I noticed that the lumber used on the project was not fully cured.

"Charles," I said. "You're using green lumber on your apartment buildings. From where are you buying your lumber?"

"Thanks for noticing," said Charles. "I need to be more careful. You see, I have my own lumber mill. When I pick up the lumber, sometimes I forget and use green lumber. I will be more careful in the future."

DAD AND I ARE PARTNERS

Charles completed construction of his apartment complex on time and below budget. He also paid every invoice on time and in full. We became good friends and worked on many projects together. I learned valuable lessons working with Charles. Perhaps the most important was never to judge another by outward appearances.

SNOW SKIING IN VERMONT
Charles Miller was my friend and a client I worked with for many years. One early winter morning, Charles came to see me about a trip to Vermont. He had a cabin there and said we could have a good time snow-skiing. I told him I had to work and didn't have time to take a vacation. Besides, I told him, I had never snow-skied. He still insisted. I still said "No."

Then he asked if I'd be willing to do some work there. I said I would if I had a job, but since there was no job, I didn't have time to go. That was on a Thursday afternoon. The next day, Charles came into the office to tell me that he needed some survey work done and expected me to go with him to Ludlow, Vermont, to do the work. He said he would have his plane ready to go early Monday morning. I thought he was kidding until he called Sunday night to tell me he would pick me up at my office around seven Monday morning.

Seven o'clock Monday morning, he was at my office, and, an hour later, we took off for Ludlow, Vermont. We flew through some snow and bad weather, but somehow, we arrived safely in Vermont. I spent the next few days surveying his property. Those days were uneventful. He then insisted I

117

go snow-skiing with him on Okemo Mountain. Okemo Mountain was one of the areas where Olympic skiers trained for the winter Olympics.

Charles took me up to what he called a small ski slope and helped me put on a pair of snow skis. It looked simple enough. Others were gliding along down the slope, with apparently not a care in the world. I put on the skis.

I had spent many days water-skiing in South Georgia and always leaned forward as I put on my skis. That was not the thing to do with snow skis. As I leaned forward, I began to move. Suddenly, I was moving forward faster and faster. I was heading down the slope. I had never snow-skied before. I'd never had any skiing instruction, and I had no idea how to turn to one side or the other on a pair of snow skis. A woman was standing a few yards in front of me.

"Look out," I yelled at the top of my lungs. "I don't know how to turn, and if you don't move, I am going to hit you."

The woman didn't move. I hit her square on her back. She went down face first. I straddled her with my skis and continued down the slope.

"Sorry," I yelled back over my shoulder. "I have never skied before and don't know how to stop or turn."

The woman looked up and smiled. "You warned me," she said.

My adventure was not over yet. A tree loomed in the distance, directly in my path. I didn't know how to turn, and my speed was approaching forty-five or fifty miles per hour. I did the only thing I could think of. I just sat down. Luckily, I came to an abrupt stop inches before the tree. I was

DAD AND I ARE PARTNERS

safe, unharmed, and a prime source of amusement for those observing my misadventures on the slopes.

Charles and I flew home the next day. I can now state, with total truthfulness, that I skied on the same ski slopes as Olympic champions. Of course, those who know the whole story, as those of you reading this book know, realize that I sat down on the job.

ALWAYS BLESS YOUR FOOD

It was summer, sometime in the early 1970s. My family and I were attending the annual meeting of the Georgia Association of Registered Professional Land Surveyors in Savannah, Georgia. Meetings were long but informative. After the meeting, many of us decided to enjoy dinner at one of the famous restaurants in Savannah. Most of us took our families with us. I took my wife, Fonda, my daughter, Mitzi, and my son, Rick.

Seafood was a specialty in Savannah, and I ordered shrimp. The servers took individual orders for everyone and delivered them to the chefs. We all waited as they prepared our orders.

About twenty minutes later, the servers brought our food. It smelled delicious and tasted even better—that is, until my son, Rick, yelled at the top of his lungs.

"Wait! Don't eat that! You'll get indigestion!"

"Quiet, son," I said.

"I won't be quiet. If the people in this restaurant eat this food, they'll all get indigestion."

By now, Rick had the attention of everyone in the restaurant as well as all the servers and the manager.

The manager came to our table.

119

"What's the matter, young man?'

"If the people eat this food, they'll all get indigestion."

"I'm sure we prepared the food well, and we have a very clean restaurant."

My wife, Fonda, and I wanted to disappear, but there was no place to hide.

"No one blessed the food, and if you don't bless the food, you'll get indigestion," said my son.

"Well," said the manager, "would you bless it?"

Rick folded his little arms, bowed his head, and gave a short blessing on the food. "The food is safe, now," he said. "Everyone can eat."

The manager was relieved. Fonda and I were somewhat relieved, and everyone went back to eating after a few laughs. It has been more than fifty years since that experience, and Rick started a tradition that, as far as I can tell, has never been broken. Before every meal, someone blesses the food.

GEORGE MELTON AND PRAYER

A young man named George Melton worked for me for a short time. He was a pleasant fellow and usually quiet. He worked on one of the survey crews.

One morning, I sent George out with a three-man survey crew on an unusually hot day. The morning temperature was in the nineties, and the temperature rose to more than one hundred degrees by noon. The work was strenuous and tempers short.

I never learned what started the argument, but two members of the survey crew had a violent disagreement. So violent, in fact, that one crew member balled up his fist and began threatening the other. George stepped between the two men.

"This is no time to start a fight," he told them. "Let's just get back to work."

That did no good. Tempers flared, and they both spoke words not fit to put in print. That was when George knelt down on the ground in prayer. He motioned the others to do the same. Astonished, the other two crew members dropped to their knees.

George started praying. After a short prayer, he looked up. The other crew members did not seem to be following his prayer. He prayed again. This time, he prayed a little longer. After that prayer, all seemed calm.

"Now let's get back to work," said George.

They did. That is the first and only time I ever heard of three rough surveyors who were arguing and about to fight kneeling in the middle of the woods to pray. But it worked. They finished the job and returned to the office as friends.

ONE-LEGGED, ONE-EYED DOG

"Don't Mess with Karate Expert Mickey Marbury or His One-legged, One-eyed Dog, Bandi."

That was the headline in a newspaper article shortly after my dad punched the editor of a local weekly newspaper on the jaw, sending him sprawling across the room.

The unusual thing about this story was that the article went on to praise my dad and demean the newspaper editor. I don't remember all the details, but I do remember some major facts.

It was morning. The newspaper editor walked into our office at 2330 Whispering Pines Road in Albany, Georgia. It was a small office, constructed of concrete block, with a large glass front door. Our secretary served as the receptionist and greeted the editor as he entered. He wanted to talk to my father.

I don't remember the conversation, but the results remain a vivid memory in my mind. After a few choice words by the editor, my dad said nothing. A few more words by the editor still brought no response from my dad. Then, with no warning, my dad drew back his fist, landed a firm blow across the jaw of the editor, and the editor hit the floor.

Dad had a small pet dog named Bandi. A few months earlier, Bandi had been hit by a car when he ran into a city street. As a result, Bandi lost one leg and one eye. Dad usually took him to the office each day. This was one of the days Bandi was in the office.

As soon as Bandi saw what happened, he hobbled over to the editor and began licking him in the face. The editor struggled to his feet and left the office. The next day, his newspaper headline read, "Don't Mess with Karate Expert Mickey Marbury or His One-legged, One-eyed Dog, Bandi." The article went on to give a glowing report about my dad and how the editor was "out of line" for his remarks.

I saw the editor a few days later. Many called his newspaper the "Scandal Sheet," because most articles were about local gossip. I couldn't understand why he'd written such a glowing article about my dad. His answer was classic.

"Your dad is one of the most loved people in this town. I am one of the most disliked people in town. I am in the

DAD AND I ARE PARTNERS

business of selling newspapers. When the most disliked person gets beat up by the most liked person, it sells newspapers. I sold more newspapers with the story of your dad beating me up than in the entire history of this newspaper."

This editor continued to write positive stories about my dad for years into the future.

NANCY CARTMELL, KARATE BOOKKEEPER

Nancy Cartmell worked with me for many years. She was more than a bookkeeper. She was a dear friend. Most surveying and engineering offices have a bookkeeper. Few have a bookkeeper who is outstanding in both bookkeeping and karate. That was Nancy. She did her book keeping well. She also instructed karate in the evenings. She was a second-degree black belt.

I had a lot of fun with Nancy and her karate abilities. The local television station once interviewed her about her involvement with karate. The moderator of the station asked several questions and then popped the attention-getter.

"Why do you study karate?" questioned the moderator.

"So I can beat up my boss," she answered.

That evening, after the TV show, I probably received twelve phone calls chiding me about Nancy and her television interview.

Nancy had a son named Chris, who was about the same age as my son, Rick. They often played together, and, sometimes, when they got into an argument, Chris would say to Rick, "My mom can beat up your dad." He was right.

One day a rather rude client entered the office. Nancy was in the front room. She served as a receptionist along with

123

her bookkeeping duties. The client approached Nancy complaining in a lewd manner that our project was not completed as soon as he wished. The more Nancy tried to explain, the more agitated the client became, resorting to loud outbursts of profanity.

I heard the outbursts and walked into the room to see if I could be a calming influence. Nancy saw me. She Looked at me and then looked back at the client.

"Ritchey," she said. "May I have your permission to beat all the profanity out of this rude jerk? I will try not to damage your office when I kick his teeth out."

The man looked at her in astonishment and then back at me.

"If he doesn't apologize immediately, you have my permission," I said.

Then I looked at the man.

"Nancy is a second-degree black belt in karate and can break several boards with one blow. She will probably break your jaw while she is kicking your teeth out. I suggest you apologize. If not, Nancy, you have my permission to teach this man better manners."

The man apologized and quietly left the office. I never found out about his complaint, but the next time he returned, he was on his best behavior.

OUR BEST COLLECTION AGENCY—THE IRS

One year, Marbury Engineering Company showed a loss of more than one hundred thousand dollars on our tax returns. This triggered an IRS audit. Actually, it would have been a

DAD AND I ARE PARTNERS

very good year had we received payment for our services. Few clients, however, paid us that year. Since we kept our books on a cash basis, the lack of payments resulted in our tax returns showing a substantial loss.

Not unexpectedly, the IRS contacted us, stating that they planned a complete audit of our books. Within a few weeks, an IRS agent was on our front doorsteps. We had all of our books together and gave him a private spot in one corner of our office. We provided him with a table, a chair, and a telephone. After several hours of reviewing our books, the IRS agent agreed with our tax report. He said he agreed, however, only if he could verify that we had not been paid.

We asked how he planned to do that. He replied that he wanted a list of all of our creditors and their telephone numbers. He said that he planned to call each of them. Accordingly, we gave him the list.

The IRS agent called every single one of our creditors. He asked them two questions.

The first question was, "I am with the IRS and doing an audit on the books of Marbury Engineering Company. Their books show that you owe them this specified amount of money. Do you plan to pay it?"

The answer from every creditor was, "Yes, we owe the money, and we plan to pay it."

The second question was, "When do you plan to pay it?"

Every answer was the same, "Immediately."

By the time the IRS agent had completed all of his phone calls, he had collected more than one hundred thousand dollars of past-due invoices. He told us our books were correct, and that we owed nothing in additional taxes. He then laughed and said he would not charge us for collecting past-due invoices but probably should. He said the government would be reimbursed for his time when he completed audits of several of our creditors who were delinquent paying their bills. I do not remember any of these creditors ever being late paying future invoices.

THE ZOO AT CHEHAW

The City of Albany, Georgia, in 1974, leased one hundred unused acres of Chehaw Park to develop a zoo, now known as the Zoo at Chehaw. My office had the opportunity of doing the zoo survey. The City of Albany transferred the animals from the existing Tift Park Zoo to Chehaw between 1975 and 1977. They opened the zoo to the public in October 1977.

I had surveyed many residential and commercial developments over the years. This was my first zoo. I learned quickly not to get too close to the animals.

Sometimes, while working around the animals, I would jump at the unusual-sounding screams from animals in cages. They were in cages, however, so I thought I was safe. So did my survey-crew members.

One day, we surveyed a parcel of land near the giraffe cages. A crew member stopped to look at the giraffe. The giraffe looked back. The crew member looked more closely, staring up at the giraffe and looking him straight in the eye.

Zap! The crew member flinched and then covered one eye. That giraffe, with perfect aim, had launched a mouth full of spit directly in the crew member's eye. We all learned that

DAD AND I ARE PARTNERS

day what it meant to stay away from the animals. It meant not only to stay out of reach but also to stay out of spitting range.

I loved to visit the Zoo at Chehaw and later wrote several children's books about the animals in order to promote this outstanding resource. I published the books on Amazon as both ebooks and as softcover print books.

RICK AND THE SWIMMING POOL

When our son, Rick, was four years old, Fonda and I took him to Doublegate Country Club to learn how to swim. Within a few months, he could swim from the diving board to the side of the swimming pool, and he would actually jump off the high dive and into the water.

When Rick was five, Fonda and I took him to Jekyll Island while I attended a Surveyor's convention. Ben DeVane and I were good friends, and we attended the meeting together. Ben's wife, Mildred, and Fonda decided to walk around the area to see some of the sights. They took Rick with them, along with Mark, who was Ben and Mildred's son.

As they walked past a large swimming pool, Rick got excited. He ran to the pool, jumped in, and started to swim back to the edge. He sank. During the past year, he had forgotten how to swim. Fonda looked at Rick and then at the water. She was dressed up and didn't want to get wet diving into the water after her son. She would, but she hoped to find a better solution. She did.

"Mark," said Fonda. "You are wearing a bathing suit. Would you jump in and get Rick out of the pool? Rick jumped into the pool and has forgotten how to swim."

DAD AND I ARE PARTNERS

Mark quickly dove into the pool, rescued Rick, and climbed out. That was all the excitement Fonda needed that day. She took Rick back to their room, where they stayed for most of the afternoon.

PRESIDENT JIMMY CARTER

I first met Jimmy Carter was when he was governor of the State of Georgia. George Busby and I were seeking his approval for construction of the proposed Sprewell Bluff Dam near Thomaston, Georgia. George Busby and I were both licensed pilots, and we flew to Atlanta, Georgia, in George Busby's plane to meet with then-Governor Carter. Our purpose was to convince Governor Carter to approve construction of the proposed dam.

This was sometime in 1974. Governor Carter believed it would be bad for the environment and vetoed the proposed construction. Although I did not agree with his decision, history shows that it was probably good for the health of the Flint River. Later, in 1975, George Busby was elected governor of Georgia and served as an outstanding governor through 1983.

In 1977, Jimmy Carter was elected President of the United States. My home was less than one hour from Plains, Georgia, and our firm was selected to do survey work for his home in Plains. That was when I realized the tight security afforded the President of the United States and how little privacy anyone holding that office had.

Gary Harrell and I drove from our office in Albany, Georgia, to President Carter's home in Plains. Gary was an outstanding and trusted member of our staff and the one I

EARLY HOME SITE OF PRESIDENT CARTER

DAD AND I ARE PARTNERS

selected to be in charge of the survey work at that site. Most of the days, the survey went smoothly. Members of the Secret Service watched our activity but interfered in no way with our work. That is, until Gary and I were surveying property close to President Carter's home. Then things changed.

Suddenly the Secret Service went on high alert. Every member of the Secret Service drew their weapons, and all of our crew was ordered to stay perfectly still. We were not allowed to pick up any equipment, put down any equipment, or move any parts of our body more than six or seven inches. Gary and I were not sure we were even allowed to breathe. This continued for three or four minutes.

Finally, the Secret Service member closest to us said, "Authorized entry—you may proceed with your work."

"What's going on?" I asked the Secret Service person standing next to me.

"Every door in the president's home is monitored," he said. "Someone entered the bathroom, and until we could verify who it was, we could not allow anyone to move."

I appreciated President Carter selecting me to do surveying and engineering work for him. I'm not sure if this statement is true or not, but someone close to President Carter told me they once asked him why he chose me to do this work. They said they told the president they understood that I had not voted for him.

The answer, I was told, was that President Carter did not make his decision on whether or not I voted for him. He made his decision based on who he believed was the best surveyor and engineer. I always appreciated that. Some may agree or disagree with his policies and decisions, but few can doubt his integrity.

FUN WORKING WITH MY DAD

The years between 1965 and 1978 were some of the best working years of my life. Why? Because those were the years my dad and I worked closest together. We surveyed together. We flew airplanes together. We designed together, and we went fishing together.

I describe many of the experiences working with my dad throughout this book. What I wish to describe here is the warmth and closeness that came from working together. The memories of those experiences last a lifetime, and, today, many years after his death, I still cherish those memories.

I remember being stuck in the mud behind an apartment complex we were designing. Dad was driving a blue Buick, and I was sitting in the passenger seat. It appeared we would have to call a construction company to rescue us. Dad insisted we continue to rock the car back and forth, back and forth, back and forth. Finally, by some unforeseen circumstance, we drove free and escaped the mud prison. That may seem minor to most, but just experiencing it with my dad made it memorable.

I remember when we built our first office building. It was in 1966, and, before that, we had worked in a pecan warehouse. Our new office building was a small, concrete-block building on Whispering Pines Road in Albany, Georgia. When we completed construction, Charles Gilbert, a friend and another member of our staff, inscribed the date—June 6, 1966.

DAD AND I ARE PARTNERS

Shortly after the contractors finished construction, I saw an advertisement in a magazine about a tree that would grow rapidly in any soil. Dad and I joked about that tree, and I ordered it. When it came, it looked like a small stick. We had just paved the alley on the east side of our office, and some cement from the soil cement base in the alley had drifted into the lawn. I decided to plant the tree in the soil cement, joking that if it would grow anywhere, it should be able to grow there. It did. Today that tree is more than thirty feet tall, and the trunk has a diameter of more than twenty-four inches.

Dad and I both earned our private pilot licenses and often flew together on business trips. Either dad would fly controls to the destination, and I would fly controls back, or I would fly controls to the destination, and dad would fly controls back. No matter when we flew, it seemed I would get the good weather, and dad would get the bad weather. There were clear skies when I flew at the controls and cloudy skies when he flew at the controls.

I remember how we had a project in Kissimmee, Florida. We flew our Cessna Cardinal to the project practically every week, and, most of the time, I flew there alone. Often thunderstorms caused us to have to land our plane before reaching Kissimmee, and the most frequent place to land was in Ocala, Florida. I landed there so much that the fixed-base operator and I became well acquainted. We visited for only short periods, but he always recognized our plane and how I would call as I approached the runway, "This is Cessna November three two two seven tango, coming in for landing."

TREE AT WHISPERING PINES ROAD OFFICE IN 2019. TREE WAS A SMALL STICK ABOUT THREE FEET LONG WHEN PLANTED IN 1966

DAD AND I ARE PARTNERS

One day, my dad flew our plane to Florida, and as usual, thunderstorms required his landing in Ocala. My dad and I looked much alike and had the same name, Ritchey Marbury, except that he was Ritchey Marbury Jr., and I was Ritchey Marbury III.

Dad approached the airport and called the fixed-base operator with the words, "This is Cessna November three two two seven tango, coming in for landing."

Dad landed, departed the plane and looked at the fixed-base operator, who had a puzzled look on his face.

"You're Mr. Marbury, aren't you? Mr. Ritchey Marbury?" the operator said.

"Yes," replied dad. "These thunderstorms sure make an old man out of you."

The operator looked again at dad and replied, "They sure do."

We never told him we were father and son. Dad and I laughed about that for years.

To me, my dad was amazing. We often checked each other's designs to be sure one of us hadn't made a mistake. I would take meticulous care with hand calculations that totaled several pages. We did not have computers in those days. Dad often did his calculations in one or two lines. We usually arrived at the same answer, but dad would do it so quickly and so simply that I always wondered how he did it.

He explained that the simple solution was usually the best solution. You can make work complicated, or you can make it simple, he would tell me. The simpler you make it, the better chance you have of arriving at the best answer. I continue striving to follow that rule.

The last major project dad and I worked on together was the Albany Mall. I have a section about the Albany Mall earlier in this book. I will describe here how Dad and I worked on the design together.

Dad asked me to do the storm-drainage design. I worked on the calculations for several days and came up with an answer. The solution was simple. We would construct a large outfall storm-drainage pipe at the northeast end of the property. It would run from the Mall property, across Ledo Road, and into Lee County. The pipe would be large, but it would do the job.

Dad looked at my solution and suggested another. His solution resulted in an outfall pipe almost half the size as my original design. Dad had a unique ability to find cost-effective solutions to every problem.

Dad's solution caused runoff from rainfall to first flow away from the outfall pipe. After providing drainage for the mall buildings, the storm-drainage pipes would eventually flow back into the outfall pipe. That way, it would take longer for the runoff from rainfall to reach the outfall pipe, resulting in a smaller pipe size. For those reading this book who are not surveyors or engineers, the longer it takes for runoff from rainfall to reach pipes, the smaller the required pipe size to carry the flow. This time is called "the time of concentration." By increasing the time of concentration, dad reduced the drainage construction cost by almost half.

I miss my dad, and I miss the many years we worked together. I am blessed to have worked with my dad, my granddad, and my son in the surveying and engineering profession. Few people have those blessings. I am grateful for them every day.

IDAHO MISSION

I interrupted my surveying and engineering business from July 1978 through June 1981 to serve a mission for my church, the Church of Jesus Christ of Latter-day Saints. Since this book focuses on my surveying and engineering experiences, I will not discuss my mission experiences here. I will say that those years were wonderful years, full of faith-promoting experiences and lifetime friendships. I served as mission president of the Idaho Pocatello Mission my first year. The mission home then moved from Pocatello, Idaho, to Boise, Idaho, where I served the last two years of my mission. The name of the mission changed from the Idaho Pocatello Mission to the Idaho Boise Mission. My wife, Fonda, served as my missionary companion.

Several things happened in my hometown of Albany, Georgia, during my mission. On September 26, 1979, the City of Albany, Georgia, conveyed the site for a proposed civic center to the Albany Dougherty Inner City Authority (ADICA). ADICA then proceeded to issue bonds for civic center construction. Later, after I returned home from my mission, I was awarded the contract for the site design of that complex. It opened February 11, 1983.

Just before leaving for my three-year mission, I obtained my Idaho professional engineering license. I never used my license during those three years, since I fully committed myself to only those things pertaining to my missionary service. The license did prove useful in later years, however, when I maintained an office in Idaho for a short time.

My dad became seriously ill during those years. He became so ill that he contracted with another engineering firm to purchase our company. At the last minute, he changed his mind. I returned home to a company deep in debt but still operational. I am grateful to my dad for his sacrifice in maintaining Marbury Engineering Company until my return home.

Chapter 5

SURVIVAL AFTER DAD'S RETIREMENT

STARTING OVER WITH NO MONEY

THE FIRST WEEK IN JULY 1981, I returned from serving a three-year mission with my church. I drove down the Liberty Expressway, observing changes made in my hometown of Albany, Georgia. Fonda and my two children could see the excitement on my face as I observed that so many of the improvements made were ones I had recommended and even prepared concept plans for before my departure.

I had rented out my home during those three years, and various friends kept my furniture. You can imagine my astonishment when we'd arrived home, opened the door, and saw that most of our furniture was back in our home in almost the same location as when we lived there three years earlier. Jan and Roy Golden, wonderful friends, had contacted those who were keeping our furniture and moved it back into our home so that it would be there when we arrived. What a wonderful way to start our new life back in our hometown.

All was not wonderful, however. We were broke. We had no money. My dad, who was my partner for so many years, was sick and unable to keep the business financially successful. The company was deeply in debt, and we had no significant backlog of business.

I needed money and some bank to lend it to me. Most banks refused. They said I had no collateral, so they could not lend me any money. Finally, I went to see Fred Sumter at First Federal Bank in Albany. He and other bank officers discussed my situation and agreed to lend me the money, even though I had no collateral.

"I hate to question a good thing," I said. "But why are you generous about lending me the money when I have no collateral and very few business contracts?"

"Well," said Fred, "we have lent money to many crooks, who did not pay us. We decided we would rather lend money to a poor honest man than a rich crooked one."

"Then, would you make sure the note says I can pay the loan back early with no penalty?" I asked. I was really kidding, as I felt I had no chance of paying the loan back early but just wanted to have some fun with my friend.

"Sure," said Fred.

One year later, I paid the loan back in full. Many times, blessings come when we least expect them.

SURVIVAL AFTER DAD'S RETIREMENT

ARCHWOOD DRIVE

Shortly after returning from my mission in Idaho, John Gay contracted with our firm to design a road from Stuart Avenue to Ledo Road. We later named the road Archwood Drive.

My son, Rick, was just learning to operate survey equipment, and he served as party chief for this project. I would go with him in the mornings to orient him, check with him around noon, and then check with him later in the day.

Rick's job was to do the field survey and topo of the property. He would then bring his field notes back to the office to be drawn and designed. As was usual with many projects in South Georgia, snakes often interrupted the work. Rick, or someone on his crew, would stop measuring, kill the snake, and then go back to work. It is amazing that, although no one wore snake protection other than boots, I never had any member of my crew bitten by a snake during my entire career. At least, not so far. I am still surveying at age eighty-one.

I had a hernia operation during the time I was designing Archwood Drive. I had to spend several days in the hospital. My family brought a computer to my room in the hospital, and I completed most of the design of Archwood Drive from my hospital bed at Phoebe Putney Memorial Hospital. By that time, most design calculations were done by computer rather than by hand, as they had been done during my earlier years.

While surveying Archwood Drive, I discovered that a small strip of land at the beginning of the project did not belong to John Gay. It was only a few feet wide, but it was owned by another party. I noted that fact on my survey drawing

141

and informed John. John assured me he would do what was necessary to gain title to the property and instructed me to continue with the design. I did.

I also designed the sanitary sewer in the road based on using plastic (PVC) pipe rather than vitrified clay pipe (VCP), which was then the required pipe material. When told by City of Albany officials to change the material from PVC to VCP, John asked my opinion about which material was best. I replied that in my opinion PVC was best but that the City of Albany wouldn't approve it.

John instructed me to continue with PVC pipe in the design, and he would discuss with the City. He also insisted that the contractor construct the sanitary sewer with PVC pipe. We were fortunate that the City reluctantly agreed to use PVC pipe as a pilot program. PVC pipe is now the standard for all sanitary-sewer-pipe construction in Albany, Georgia.

John was not so fortunate with the property encroachment. The property was not his, and even though only a few feet were involved, the owner of the encroached property said John did not gain his permission to build a road across his property and sued for damages. John lost the lawsuit.

John later told me that he felt he would lose the lawsuit but that this way was still less expensive than if he were not able to build the road where he wanted it located. I always enjoyed working with John Gay, but sometimes his methods were unorthodox.

On completion of the road, I tested the final construction for compaction. A heavy dump truck filled with sand drove over the road while my son and I walked along behind,

SURVIVAL AFTER DAD'S RETIREMENT

looking for any deflection in the pavement. Although that testing method is sometimes used today, usually tests today are made by soil-testing laboratories.

Oxford Construction built the road, and I found no flaws in the construction. After final construction approval, I prepared a final survey of each lot, recorded the plat, and finalized the project.

PECAN STREET—CORDELE, GEORGIA

Jean Burnette was city manager in Cordele, Georgia. She was a hands-on manager with a reputation for wanting everything to be done precisely and done her way. The City of Cordele selected our company to do the surveying and engineering work on that part of Pecan Street south of Sixteenth Avenue, with one stipulation: we had to design the street so that all construction would be within the existing right-of-way. The street already existed as a paved street with ditch sections. Our job was to design the street with a wider paved section and with the addition of curb and gutter.

I discussed the project with Rodney Hutchinson, an outstanding designer with our company. I believed it could be done, but I wanted another opinion. Rodney agreed with me. We both believed we could design the project so that all construction would stay within the existing right-of-way. We accepted the project, and I put Rodney in charge of the design. In his usual professional manner, Rodney designed the project to meet all requirements.

Even though the design went well and the required approvals were quickly granted, construction was a greater challenge.

143

Property owners along the street grew upset with every shovel of dirt. Mrs. Burnette received angry phone calls telling her how the construction was ruining their property values. Although all disturbed shrubs were in the existing right-of-way, property owners considered them their personal shrubs and expressed displeasure whenever construction damaged any shrubbery.

We held many meetings with angry property owners to reassure them that the completed project would be an asset to everyone living on the street. My staff and I spent many hours in excess of those budgeted to meet every possible demand from angry property owners.

The contractors completed the project on time and within budget. After completion, many property owners called Mrs. Burnette expressing thanks for the project. Some even apologized for their actions during construction. Many of the property owners were the same ones who'd expressed the greatest anger during construction. After project closeout and final inspections, I presented my last invoice to Mrs. Burnette.

Mrs. Burnette looked at the invoice and said, "Ritchey, you must have spent much more on this project than you billed us."

"We sure did," I replied, "but I invoiced you only for what was in our contract."

"Well," said Mrs. Burnette, "I know you and your staff worked many extra hours to make this project meet all our requirements and keep the adjacent property owners happy at the same time. Send me another bill for your additional cost, and we'll pay it. It's good to have an engineer and surveyor who puts our needs ahead of their own."

I sent the bill for the additional cost, and the city paid it promptly.

PEANUT HULLS IN THE DITCH

John Gay and I often laughed about how I instructed him to backfill a thirty-foot-deep ditch with peanut hulls. John was the sewer contractor, and I was the engineer in charge of the project. I had my engineering degree from Georgia Tech. I had a master's degree from the same university. I had dual registrations as both a registered land surveyor and as a registered professional engineer. My college textbooks said this would work. Even with being out of college for about twenty years, I still believed things always worked exactly as shown in textbooks. What was I thinking?

The project location was just east of a new civic center under construction in Albany, Georgia. The proposed sanitary sewer would be located along the center of a paved street. My school textbooks explained that a backfill method called the "imperfect trench" would prevent the need to encase deep sewers in concrete, which was the standard method of preventing deep sewers from collapsing under the heavy weight of soil over the sewers.

The volume of soil one foot wide by one foot long and thirty feet deep is thirty cubic feet. The weight of the soil that would be over the sanitary sewer was about one hundred ten pounds per cubic foot. The weight of just one-foot-wide section of soil one foot in length and thirty feet deep would be about three thousand three hundred pounds, or a little more than one and one-half tons. Of course, the soil has some bridging effect, but still, that weight is enough to crush a sanitary sewer pipe unless given some additional protection.

In theory, an "imperfect trench" would cause the soil to spread its weight away from the pipe, preventing it from crushing. At least that is what the textbooks said. Peanut hulls were available for free, and encasing the pipe in concrete was a huge expense.

John told me I was crazy. I told him I was sure it would work because my college textbooks said it would. What I did not consider—something that I would consider after many more years of experience—was that this would work only if the contractor did the work perfectly. Even though it was then more than twenty years after my graduation, I still failed to consider that contractors do not always do things exactly to specifications.

I was fortunate. John followed specifications more closely than he ever did on any other project. Several years later, floodwaters inundated the street. They also flooded the parking lot just west of the street and a civic center that had been constructed at the same time as the sanitary sewer. The parking lot had settlement in several places due to the floodwaters. The civic center had settlement due to the floodwaters. The street with the sanitary sewer covered by the "imperfect trench" did not settle.

I am older and more experienced now. Today, I would not be so sure of myself. I still understand the theory and believe that if the contractor does everything perfectly, it will work. I also understand how few times even the best contractors do everything perfectly. I also understand how easy it is for engineers and surveyors like me to make mistakes, no matter how careful we are. Fortunately, for me, this was one of those times that everything went right. I would not use that method again, however. It's too easy for too many things to go wrong.

MUDDY MOTIVATION

Motivation is an interesting thing. It was a muggy summer day in July. Our job was to survey property in the Percosin swamp, located in southwest Georgia. I was the party chief and also the instrument man. I carried a geodimeter to the site. It was a distance-measuring device that weighed about forty pounds.

By then, most surveyors made measurements with distance-measuring devices. They were faster and more accurate than the older methods of surveying with chains and chaining pins.

We waded into the swamp. Walking was hard, and mud covered our ankles. Every day, we saw a moccasin, a rattlesnake, or both. Every day, mosquitos enjoyed taking some of our blood. Some days, we even had to change directions in order to avoid alligators, which also made their home in the swamp. No one wanted to be there, and morale was growing worse with every step.

Then it happened. I stepped into a hole. I struggled to keep my balance, but it was useless. I fell, face first, into the mud. The forty-pound geodimeter felt like a giant press pushing me deeper into the mud.

I tried to get up, but every time I tried, I fell again. A mudpack covered my face, and it was not for beauty. Finally, one of my crew reached out his hand and pulled me up. About halfway up, he threatened to let me go, but he didn't.

Laughingly, he remarked, "Do that again. I haven't had so much fun all day."

We completed the survey in about a week. One crew member said that, if I did not increase his pay, he would push me into the mud again. I told him I needed the money to pay cleaning expenses, so I could not increase his pay until the end of the year. He just laughed and said seeing me all muddy was pay enough.

WATCH THAT MACHETE—DON'T SCALP MY SON

Shortly before noon on one quiet, sunny day, the hospital called for Mr. Ritchey Marbury. I answered. The voice on the other end of the line said that my son, Rick, had been cut in the head with a machete and was receiving treatment at the hospital emergency room. They said I should come immediately.

I called my wife and rushed to the hospital. As is usual with emergency rooms, it took a little time for them to let me in. Sometimes emergency rooms are not the epitome of efficiency.

I entered the emergency room and saw doctors rushing around and going in and out of closed cubicles. The doctors were all dressed in white, with face masks covering their mouths. Nurses followed the doctors with serious expressions but saying nothing. I waited nervously. I envisioned my son with the top of his head scalped and near death.

Finally, I saw my son. I rushed to his bedside. Blood covered the sheet that lay across his body. Blood covered his head, and some blood was still on his face. I couldn't even see his face at first. I was in shock from all the blood-covered sheets. Finally, I looked more closely at my son's face. He was smiling.

"Head cuts bleed a lot," explained the nurse. "The cut on your son's head is not serious, and he's fine—but you're not!"

An attendant quickly brought me a wheelchair. She told me to sit down and put my head between my legs. All the blood had drained from my face, and I was as white as a clean sheet. I was near fainting. When my son hurt, I hurt.

When my wife arrived, she was much more composed than I was, although still distressed. Rick recovered quickly and soon returned to work.

CERTIFY PLAT NOT CORRECT

A problem I often had with reviewing agencies and attorneys was the certifications required on plats. Sometimes they were proper, such as certifying the plat closure or the type of equipment used on the survey. Other times, the requirements were ridiculous or so extreme that they required us to certify that our drawing depicted everything existing not only above ground but also below ground, even when no underground exploration was conducted.

I wonder if these characters believed every surveyor had X-ray vision and could automatically tell the location of any utility above or below ground and within fifty feet of the site. This was the case with a survey I made in a small town in

South Georgia. I will not name the town or the subdivision, but this is the story of an actual survey and plat that a fellow surveyor, Jim Faircloth, and I prepared many years ago.

We had just completed the final survey and plat for a small subdivision. As requested, we submitted the plat to the planning commission for final approval. All corner markers were located and described in the plat. Street names and street right-of-ways were clearly marked. We conducted a thorough review of all checklist requirements and verified we complied with them all. We included the required certifications, with one exception. The planning commission required we certify to the location of all above-ground and underground utilities. We located all above-ground utilities but had no idea of the location of many underground utilities.

Searching the city and county records, we found plats showing approximate locations of underground power, telephone, cable TV, and other utilities. We felt sure even more underground utilities existed that were not even listed in existing records. We explained this to the approval agencies, stating that we couldn't certify to the exact location of all underground utilities. That didn't matter. The agencies said their checklist required a certification regarding all above and below ground utilities, and the plat would not be approved without such certification.

We were in a dilemma. If we certified all underground utilities were shown and correctly located, we risked liability for certifying to incorrect information. If we failed to include the certification, the plat would not be recorded. After much deliberation, we reached a conclusion. The requirements were

SURVIVAL AFTER DAD'S RETIREMENT

that we include a certification regarding all utility locations. The requirements did not state that we certify that all utility locations were shown correctly.

I don't remember the exact words we used on the plat, but they were something like this: "We certify that the above ground utilities are shown in their actual locations. Approximate locations of underground utilities are shown as indicated on plats provided us by others. We certify that other underground utilities are believed to exist of which we have no knowledge and, therefore, are not shown on the plat."

We presented this certification to the reviewing authorities explaining that this was, indeed, a certification regarding all utility locations. They accepted the certification and recorded the plat.

PIT TOILETS IN THE NEIGHBORHOOD

Sometimes reviewing agencies are easy to work with. Sometimes they're not. I'll describe here one of those "not easy to work with" times.

It all began with a simple request to put a septic tank on a one-acre lot in an upscale neighborhood. Most of the neighborhood was served by septic tanks, since there was no easy access to sanitary sewer. There was sanitary sewer some three hundred feet away, but it would be necessary to construct the sewer through a swimming pool in order to reach the property with a sewer only three hundred feet long.

The other way to gain access to sanitary sewer was to construct more than one thousand feet of sewer along street right-of-way until it reached the lot under consideration. This was a costly alternative, and since the other homes surrounding the lot were served by septic tank, we thought this lot should be allowed to be served that way, also.

We were wrong. When we applied for septic tank service, the health department rejected the application.

"You have sanitary sewer within five hundred feet of your lot; therefore, you must serve the lot with sanitary sewer. Our regulations require that all lots within five hundred feet of an existing sanitary sewer line must be served by sanitary sewer."

I tried, to no avail, to explain that sanitary sewer was really available only more than one thousand feet away, since constructing a sewer line through a swimming pool was unrealistic. The health department still insisted the lot be served by sanitary sewer.

I looked at the cost of constructing the sanitary sewer along the street. It could be done, but the cost would probably exceed twenty thousand dollars. The health department still insisted that the rules be followed and the lot be served with sanitary sewer. I requested a variance, but that was denied.

I studied the department regulations to see if there were other alternatives. Surprisingly, there was one—not really a reasonable one but one that was approved by their regulations. The manual clearly described how to design and construct a pit toilet. Not only that, a pit toilet was allowed by ordinance in the neighborhood.

My client had a reputation for doing rash things on occasion, and I told him what I'd found. I said I would like to design and submit a pit toilet for use on his lot. Of course, we all knew he would never do that, but we liked the idea of having some fun with the reviewing authorities.

I prepared full-scale drawings of a pit toilet, using every requirement in the health department regulations. Then I submitted the plans.

At first, the health department thought it was a joke. Actually, it was, but my client and I kept a straight face as we told them the cost for sanitary sewer was far beyond our budget. Since my client had committed to build the home and could not afford the extra twenty thousand dollars to bring sanitary sewer to the lot, the only alternative left was to construct a pit toilet.

We again requested a septic tank variance. It was approved.

PURCHASED STEVENSON AND PALMER

I have made many good and many bad business decisions. Thankfully, most of my business decisions were good ones, and that's why I was able to stay in business. One of my worst business decisions, however, was the purchase of Stevenson and Palmer Engineering in the late 1980s. They were a great company. The owners were men of high integrity and professional competence. My mistake was that the purchase was a leveraged buyout. I borrowed more money than I could afford to borrow and used that to purchase the company.

In all fairness, Joe Palmer, one of the owners of the purchased company, warned me that they had a unique way of marketing services. They would prepare preliminary plans at no cost, with the understanding that all costs and services would be paid for at the time financing was obtained for the projects. Most of their projects were in small communities. Joe Palmer was a master at obtaining financing and, therefore, would do the initial work on the project at no cost. Once the project obtained financing, the communities signed contracts with Stevenson and Palmer to complete the project. The cost of the preliminary plans was included within the signed contract.

This still could have been a worthwhile arrangement except for one fact. Within one or two months of purchasing the company, most funds to communities were frozen for an unspecified time. This left me with large payrolls, a large backlog of work, and no way to get paid in the immediate future. After a year, the original owners of Stevenson and Palmer agreed to take back their company and complete the contracts. Although this relieved me of any debt to Stevenson and Palmer, I was still heavily in debt to local banks.

One irony of this situation is that, within a few months of Stevenson and Palmer taking back their company, funds to finance local communities were unfrozen, and Stevenson and Palmer received complete payment for all of the work done previously.

Chapter 6

MARBURY ENGINEERING BEGINS AGAIN

UP FROM THE ASHES

THE EVENTS FOLLOWING THE DISASTER of my previous unwise business decision make me reflect on one of the sayings from the Walt Disney production, *Chitty Chitty Bang Bang.*

"Up from the ashes. Up from the ashes. From the ashes of disaster come the roses of success."

Knowing that I was heavily in debt, most of the local banks refused to lend me any money. If I could not get paid quickly for all surveying or engineering work I did for clients, I would have zero cash flow and not be able to survive.

I will always be grateful for the events that followed. I felt I needed to be honest with my staff. I brought them all into a room together. I explained to them my financial situation, which they probably already knew, and said it would be necessary to make a significant adjustment in all of their salaries. They nodded their heads, and all said they understood.

"No, I don't believe you do understand," I replied. "All of you are going to receive a significant *increase* in salary. In order to survive, we will all have to work more efficiently and effectively. If we are successful, each of you will have earned your increase in salary. If we are not successful, I will go out of business, and being a little more in debt won't make any difference."

"How will you manage to pay the increase in salary?" asked one of the staff members.

"I will cut my own salary to practically nothing and hope to live on borrowed money. I will do this until our company manages to get back to a positive cash flow."

I will always be amazed at the results of that year. Thanks to a superb effort on the part of our staff, after one year, our company showed a positive cash flow. We were never late with a payroll, and we never missed a payment to the bank.

HIDDEN LAKES

Bob Barkley walked into my office sometime in the late 1980s. He owned several hundred acres of land in Dougherty County, Georgia, just north of Gillionville Road. The property was beautiful, and he proposed to develop a subdivision consisting of several lakes. Luxury homes would surround the lakes.

He asked me to give him a price to prepare a preliminary layout and do a complete design of the subdivision. When I gave him a price proposal, he told me he had a better price from another engineer.

MARBURY ENGINEERING BEGINS AGAIN

**HIDDEN LAKES
ALBANY, GEORGIA**

I told him that I understood but would like to offer a suggestion. I told him I would prepare a preliminary layout and let him evaluate it. If he chose to use my layout, he would pay the price in my original proposal. If he chose not to use my layout, the work I did would cost him nothing. He agreed.

I prepared the layout. He liked it, and that began many years of friendship and many years of my providing him with land surveying and consulting engineering services.

He chose to name the subdivision "Hidden Lakes." True to his word, Bob paid every invoice on time and without question. What I really liked about working with Bob was his integrity and his commitment to quality. The lake certainly made the subdivision attractive and valuable. His creativity made the development even more valuable.

Toward the northeast part of the property, I proposed two lakes. The lakes would be joined by a road which crossed a large box culvert. The box culvert allowed water to flow from the proposed lake on the northeast side to the proposed lake on the southwest side. Bob wanted that culvert to look more attractive than simply a box culvert. His idea was to make the culvert look like a covered bridge.

Across the top of the culvert, Bob had a structure constructed that looked exactly like a covered bridge. He also wanted those who drove over the culvert to feel that they were driving over a covered bridge. In order to do this, he had railroad ties placed across the road and asphalt paving placed around the railroad ties. This gave a bouncing effect to all who crossed the culvert, thus creating the feeling that they were crossing under a covered bridge.

MARBURY ENGINEERING BEGINS AGAIN

In order to keep the lakes holding water during dry periods, it was necessary to construct a well and pump. The well would pump water into the lakes during the dry season. Bob was agreeable but didn't want residents to see a pump in front of the lakes. He solved this problem by constructing a pump house around the pump. The pump house looked like an old-fashioned outhouse.

Bob wanted additional features around the lake on the northeastern side. To accomplish this, he constructed a walkway from one of the roads to a point near the center of the lake. There he developed a small park that could be used by residents within the subdivision.

Bob placed light-blue dye in the lake waters to give the lakes a more aesthetic appearance. Everything he did was first class, and he did everything he promised to do. I told many people that I would rather have Bob's word that he was going to do something than a signed contract prepared by some of the most competent lawyers.

During the 1990s, Dougherty County experienced two major floods. One was the flood of 1994. This flood caused little damage to Hidden Lakes Subdivision. The flood of 1998 was a different matter, however. Not all of the outfall construction had been completed in the subdivision, and rising floodwaters indicated that some of the houses downstream would experience flooding if preventative steps were not taken quickly.

Bob called me and asked what I could do. I looked at my drainage calculations and realized that water was flowing from developments more than a mile upstream from the

159

subdivision. This water would flood homes in Hidden Lakes Subdivision in about twenty-four hours.

There was a solution. One of the upper lakes had storage capacity to handle additional water. If we blocked the water from flowing out of the upper lakes, we could prevent the downstream flooding.

I called Wayne Cowart, who worked with the Dougherty County Public Works Department. I explained the situation and requested his help. Within about one hour, crews from the Dougherty County Public Works Department were onsite with equipment to do the job.

The solution worked. None of the downstream homes flooded. One of the homeowners, whose home would have been flooded if these precautions had not been taken, baked me a pie and brought it to me as a token of appreciation.

I loved working with Bob Barkley. He was always a close friend.

LANCASTER VILLAGE

Larry Walden was another client and friend. He sold me the land on which I built a new office in 1990. The location was 2334 Lake Park Drive in Albany, Georgia. It was a large, colonial-style office with brick walls and white wooden columns. It had the perfect location, and the surrounding development enhanced even more the value of my new office.

One day, Larry approached me and announced he was putting a permanent fruit stand right in front of my office. I couldn't believe it. Larry was a friend. His developments were always first class. Why would he devalue my office by placing

MARBURY ENGINEERING BEGINS AGAIN

LANCASTER VILLAGE

a fruit stand right across the street? In what seemed to be the ultimate insult, he hired me to design the development. He also indicated he was bringing in another building he'd found in Coleman, Georgia, to place in the development.

I thought, *A fruit stand in front of my beautiful colonial-style building is bad enough, but to try to save money by just moving an existing building he'd found in Coleman, Georgia, and placing it in front of my new office, rather than constructing a new building, is going too far.*

Larry didn't crack a smile when he left the office. As I worked on the site design, I learned "the rest of the story."

The development became a shopping center created to look like a small village. The fruit stand depicted an old country store. The building he moved in from Coleman, Georgia, was an old church building. After all, every village had a small church. The inside of the old church building was remodeled into an antique shop. He constructed a school-supply store in the image of a red schoolhouse with a bell on top. Other buildings in the development added to the village effect. Even the pavement gave the feeling of a small village. Although the pavement was actually asphalt, the contractors used a special process to make the street look like an old brick-paved street.

Larry named the development "Lancaster Village," after Frances Lancaster. Frances served as secretary and administrative professional for both Larry and his father for many years. My father and I also considered her a dear friend. To me, it was an honor to have my office across the street from a development named after such an outstanding lady.

CALLAWAY LAKES

In the early 1990s, Cason Callaway approached me about designing a residential development on his five-hundred-plus-acre site in Lee County, Georgia. The property was in a great location but had a number of design challenges. One of the challenges was the fact that many depressions on the property made it difficult and expensive to develop. Storm drainage design would be difficult.

Another challenge was the fact that a former Nike missile base was located on the property. We had to find a way to use or remove the Nike missile base.

My son, Rick, worked with me at that time. We decided to develop several schematic drawings showing alternative ways to develop the property. Rick worked on one proposed layout while I worked on another. After reviewing the layouts, Cason Callaway decided he liked the schematic layout prepared by my son. We immediately went to work designing the property based on Rick's schematic layout.

In order to meet the storm drainage challenges, we decided to take advantage of the depressions in the topography to create lakes. These lakes would also serve as storm drainage detention ponds. The detention ponds would store water from heavy rainfalls and release the water slowly in such a way as not to flood adjoining property. Since these ponds would be designed as lakes, they would be landscaped and become a major attraction of the development. With the lakes being a focal point of the development, we recommended the development be named "Callaway Lakes."

At first, Cason was reluctant to use his name as the name of the development. We reminded him, however, that "Callaway Gardens" was a famous resort in Georgia and that the name "Callaway" would add much value to the development. With this understanding, he agreed to allow us to name the development "Callaway Lakes."

Next, we turned our attention to the problem of the abandoned Nike missile base. The first plan was to demolish the base. The Nike missile base was similar to an underground cave, except this cave had strong, reinforced concrete walls. The walls were so strong that it was practically impossible to demolish them.

After discussing solutions with several contractors, one contractor suggested filling the inside of the missile base with earth. He also suggested covering the top of the base with earth. This would allow the area to appear as a large mound of dirt, which could be used as a park within the development. We adopted this solution. The development was an immediate success.

HURRICANE ANDREW

On August 24, 1992, Hurricane Andrew hit Homestead, Florida, as a category-five hurricane. At that time, Hurricane Andrew was the costliest in the history of the United States, costing, by some estimates, more than thirty-two billion dollars in damages in the state of Florida alone. I, along with other staff from Marbury Engineering and local churches, volunteered to drive to Homestead to help. We were told to come fully self-contained, as no hotels or other facilities that

MARBURY ENGINEERING BEGINS AGAIN

offered food or lodging were available. We were also advised to bring tire-repair kits, as flat tires were common.

We set up a campsite in the backyard of one of the Homestead residents. Much of the city was totally without buildings, street signs, trees, or any other landmarks that could indicate our location. In some areas, even the local residents had trouble finding their way because of the total devastation. It looked like a war zone.

Eventually I found my way to the city engineer's office and told him we'd come to help. I explained that we were totally self-contained and brought skilled labor, tarps to cover damaged roofs, tools, and other equipment. I said the only thing we might need was fuel for our vehicles, as most gas stations were not working. He quickly gave us access to city fuel.

When he found out I was a registered engineer, he wanted to know if I had any suggestions. I suggested that city building codes be suspended for a few months with the understanding that final repairs would meet all codes. The immediate need was temporary shelter, and most of the temporary repairs would not meet codes. I also explained that, although many of the workers were licensed in the state of Georgia, they did not have Florida licenses. Many were skilled workers, and there was no time to obtain Florida licenses and meet the immediate needs. The city engineer immediately approved all my suggestions, and we started work.

I stayed only a few days, but others stayed a month or more helping in a variety of ways. We repaired roofs, helped set up temporary tent cities, and helped arrange for food, clothing, and other necessities to be brought to Homestead. The next week for me was work as usual.

GRAND ISLAND

One of the best places in southwest Georgia to hunt doves was a site just north of Ledo Road in Lee County, Georgia. The property was planted in corn, and after the corn was harvested, doves flooded the area. Hundreds of doves flew between the cornfields and the many waterholes on the property.

The owner of the property, John Gay, approached me one day in 1993 and said he was going to develop a golf course on the property. He had selected the golf-course architect but wanted me to do the storm drainage design. When his original agreement with the golf-course architect did not work out well, he asked if I could design a golf course.

"I believe I could," I told him, "but I wouldn't hire me to design a golf course. I have never designed one."

"I still want you to do the design," said John. "Will you do it?"

"If you want me to do it, I will," I said. "I'll get some help from friends who have a great deal of experience with golf courses, and we'll get the job done."

I contacted Ray Jensen. He was a close friend and one who had worked on hundreds of golf courses. Ray was a soil scientist and an expert on turf grass. For many years, Ray Jensen helped with maintenance at the site of the Masters Golf Tournament in Augusta, Georgia. Ray agreed to help and suggested we use his good friend, Don McMillan, to do the construction work.

MARBURY ENGINEERING BEGINS AGAIN

We obtained an aerial photograph of the property, and, with the help of Ray Jensen, I completed design of the routing plan for the golf course. Although I was a poor golfer myself, I had another friend who was an outstanding golfer. His name was Kevin Meadows. I decided to have some fun with my golf-course design.

Kevin could hit the ball absolutely straight, and it would land at a point more than two hundred fifty yards from where he hit the ball. I could hit the ball approximately two hundred yards, and it would generally slice to the right.

I decided to design the seventh hole so that, if Kevin hit the ball the way he normally did, the ball would land right in the middle of a sand trap, which I designed in that specific location. At 200 yards, I designed a steep downhill slope to the right of the fairway. My idea was that, if I hit the ball the way I normally did, it would slice to the right, roll down the slope, and result in a drive of more than 250 yards.

The design went exactly as planned. The first day we played the course, Kevin hit his drive so that the ball landed in the middle of the sand trap. It buried itself deep in the sand. My drive sliced as usual, rolled down the long, steep slope on the right of the fairway, and ended up in the middle of the fairway, more than 250 yards from the tee. Kevin still parred the hole, and I ended up with a double bogey. Nevertheless, it was fun to see my design get the results I'd proposed.

The way I designed the driveway in front of the clubhouse was another interesting method. The drive went up a steep hill in front of the clubhouse and back down the hill again. My original design consisted of a series of compound curves. The calculations were elaborate, and I was quite proud of myself for what I had accomplished.

The owner, John Gay, had another idea. He had me drive my vehicle from the driveway approaching the clubhouse, up to the front door, and back to the driveway leaving the clubhouse. He then had me design the driveway based on the tracks from my vehicle. His plan worked perfectly.

The Grand Island Golf Course opened for play in 1995. It was a par seventy-two course, with a distance of 7,012 yards from the longest tee locations. It was a wonderful course and was enjoyed by many for twenty-one years. Due to financial difficulties, the course closed on December 15, 2016.

ITCHAUWAY PLANTATION

Marbury Engineering was fortunate to be selected to perform the boundary survey for many large plantations. These included a 1,580-acre boundary survey in Lee County, Georgia, and a 16,000-acre boundary survey of Senah Plantation in Dougherty County, Georgia. One of the largest boundary surveys, however, was the 28,000-acre boundary survey of Itchauway Plantation in Baker County, Georgia.

The Itchauway Plantation boundary presented all the problems associated with large boundary surveys. What if we worked for months and, on completion, found there was an error in the survey? If we had an error, would we need to perform the entire survey over again? What procedures could we use to reduce the probability of errors?

Much of the property was not accessible by our cars or trucks. How could we reduce the time it took to walk to the point where we'd left off surveying the previous day? Part of the boundary was on a creek. How could we survey the creek bed and accurately determine the center of the creek?

We located the center of the creek by surveying along the edge of the creek and placing reference points where the creek changed directions. We then measured from the reference points to the creek's center. I explained various methods of doing this in an earlier section.

There was an easy solution to the time-and-accessibility problem. Near the end of the first day, Cary Reed, the party chief in charge of the field work, suggested we purchase two four-wheelers. He suggested that the time saved by using the four-wheelers would more than pay for the cost. In addition to that, we would have the four-wheelers to use for future projects. It took me less than thirty seconds to say "Yes," and, in another few minutes, Cary and I were on our way to the store and purchased two four-wheelers.

This is one of the big advantages of a small company. We had no purchase orders to prepare. We had no board of directors from which to obtain permission. We did not have to explain why this purchase was not in the year's budget.

It did not take an accountant to understand that this purchase would save as much time and money on this project alone to pay for the two four-wheelers. The benefits were obvious, and we acted immediately. We probably saved more than double the cost of the four-wheelers on that project alone. Years later, when I worked with a larger company, I looked back with fond memories on the simplicity with which I could make common-sense decisions with a small firm.

MARBURY ENGINEERING BEGINS AGAIN

The four-wheelers solved the problem of easy access, but the problem of preventing errors still existed. In fact, our first survey around the property did not close—or, in other words, had a major error. In survey terms, we had an error of closure exceeding one foot in several hundred feet. The survey had taken many months. We didn't want to resurvey the property again; however, we obviously had a major error somewhere.

I did the mathematical calculations and could not find the error. We looked for possible transcription errors but found none. There was only one thing to do—resurvey.

We decided to survey a random line through the center of the property in order to divide the tract into separate parcels. If one parcel met survey-accuracy requirements, and if the other parcel did not, this would isolate the error. It worked.

Next, we looked at the direction of the error. I don't remember if the error was in an east-and-west direction or if it was in a north-and-south direction. I do remember that the direction of the error matched closely with one particular line on the survey.

We resurveyed that line and found we had recorded the wrong distance. On redoing the calculations, the survey met accuracy requirements. We set final corner monuments. Our drafters completed the drawing, and I placed my registered land surveyor's stamp on the final plat and delivered it to the owners.

RED AND GREEN ARE SOMETIMES BLACK AND WHITE

One of the great lessons I learned during my career as a land surveyor and civil engineer was the danger of criticizing too quickly. Jesse Glover was a fine designer and CAD operator. He worked hard and often stayed late to complete the day's goals. It was one of those stressful days. I reviewed some drawings prepared by Jesse and asked him to make a few corrections. It should have been easy to accomplish. After all, I'd marked the needed corrections in red and the areas to remain unchanged in green.

Jesse presented me his revised drawings an hour later. Some of the areas to remain unchanged were changed, and many areas requiring changes were left unchanged. In those days, I was not known for my patience and calm disposition. That was one of my bad days. Red-faced and talking in not-so-quiet tones, I asked Jesse why he had not followed instructions.

"I did," replied Jesse.

"No, you didn't. You ignored my instructions. I plainly marked the changes in red and those areas to remain the same in green. Can't you tell the difference between red and green?"

"No."

"Why?"

"I'm color blind."

In an instant, I had ruined the desire of one of my most loyal staff to do anything other than what he was required to do. He had worked late to make the changes, and I had done nothing but criticize him for something that was my fault. If I had taken the time to review with Jesse the needed changes, they would have been made. Instead, I had thrown the drawing on Jesse's desk and demanded the changes be made, with no comment other than, "Don't talk back—just do it."

MARBURY ENGINEERING BEGINS AGAIN

One important lesson I learned from that experience was never to blame before examining all the facts. Listening and understanding are always more important than blaming and rebuking.

SNAKES AND STEPHEN HART

Surveyors sometimes play practical jokes on members of the office staff, and, too often, the brunt of these jokes are secretaries and administrative assistants. Such was the case one spring morning in our Albany, Georgia, office. Stephen Hart was the perpetrator. Our secretary was the victim.

Stephen found a fake snake that looked as much like a real snake as the actual thing. He kept it in his pocket for a long time, but his natural, fun-loving instinct got the best of him. Our secretary was a very proper lady. She was the epitome of efficiency and productivity. Each morning, she arrived at work precisely at the proper time and started work immediately. She kept pad and paper in her middle desk drawer, which she opened first thing before starting on morning assignments. Stephen usually arrived fifteen to thirty minutes early.

Stephen was your typical fun-loving survey party chief. He liked the outdoors and everything about it. He loved fishing and had no fear of water moccasins, rattlesnakes, or critters of any kind. Our secretary, however, grew traumatized by even the thought of snakes. She'd mentioned that fear often enough that Stephen could not control himself that fateful spring morning.

It was about 7:45 a.m. Stephen fetched the fake snake from his pocket and carefully placed it in the middle drawer

173

of our secretary's desk. To be certain it would have the proper effect, he positioned the head against the top outside edge of the desk, making sure that the head would fall to the front of the drawer when it opened. This gave the appearance of the snake moving as the drawer opened. The trap was set, and our secretary arrived at precisely eight o'clock.

She should have noticed everyone watching intensely as she opened the desk drawer, but she didn't. Everything went according to plan. She opened the drawer. The head of the fake snake fell to the front of the drawer, and she let out a scream "heard around the world." She was normally a gentle lady, but that day, if she could have caught Stephen, he would have been tarred, feathered, and dropped into a pit of poisonous vipers.

This was not the only occasion on which Stephen Hart had fun with snakes. Willie C. Wimberly worked on Stephen's crew. He was a great worker and very loyal to Steven. He also had a fear of snakes. Someone said that, if we wanted some brush cleared quickly, just point Willie C. in the direction of the brush needing to be cleared, and place a snake behind him. He would clear the path in minutes.

One afternoon, Stephen killed a large rattlesnake and put it in the back of the work van. Willie C. was apprehensive, but since the snake was dead, he rode back in the van with Stephen. When they arrived at the office, Stephen placed the snake on the ground and told Willie C. to be sure it was dead. As Willie C. glanced at the snake, Stephen pressed hard on its tail. The snake recoiled, and its head sprang into the air. Stephen laughed. Even a dead snake will move when its tail is pressed, due to muscle spasms. Willie C. believed the snake was still alive. He disappeared with the snake's first movement.

I LOST MY DAD

On February 6, 1994, I lost my dad. Although he had retired when I returned from my Idaho mission in 1981, he had continued to be a mentor to me. We still went fishing together. We attended Rotary Club together, where we were both members, and we sometimes went for drives just to reminisce about old times and former projects. Dad developed Alzheimer's disease in his later years but always recognized me as well as Fonda and our two children, Mitzi and Rick.

It's been more than twenty-five years since I lost my dad, and I still miss him. No child could have had a better father. We worked together, played together, were business partners, and enjoyed most of life's saddest and happiest moments together. Dad always taught me that doing the right thing and doing the best job you could do were more important than making a profit. He said that, if you do your job well, profits will come. He was right.

Dad taught me that there is no excuse for not doing your best. He was always faithful to my mother and loved her with all of his heart. He taught me that our most important business was serving others, and he always lived by the Rotary motto, "Service above self. He profits most who serves best." Dad was more than just a father—he was my best friend.

Every day, my dad lived the governing values of Marbury Engineering Company, which were "Integrity, Quality, and Service."

FLOOD OF 1994

On July 3, 1994, Tropical Storm Alberto hit Florida. Maximum sustained winds reached 60 miles per hour. I did not feel any impact from the winds, but floodwaters set records in my hometown of Albany, Georgia. Albany received 6.88 inches of rain between July 3 and July 7, 1994. The city of Americus, just north of Albany, received 27.06 inches. Peak flow in the Flint River reached 120,000 cubic feet per second. That was more than thirty percent higher than the previous peak record flow in 1925 of 92,000 cubic feet per second.

Normal elevation on the Flint River is about 150 feet above sea level. On July 11, 1994, the Flint River crested at a flood stage of forty-three feet, an elevation of 193.13 feet above sea level. The highest previous flood stage was 37.8 feet, in 1925.

More than four hundred caskets from two public cemeteries floated from their grave locations. An estimated 5,802 housing units were damaged. More than nine thousand acres were flooded. Approximately twenty-four thousand residents evacuated their homes. The flood caused losses to many major businesses, including K-Mart and Dairy Queen on Radium Springs Road and Sonny's Barbeque and Piggly Wiggly on Slappey Drive.

This was a busy time for me. As the Flint River began to crest, a member of the Georgia State Board of Regents called me to ask if I thought they should evacuate Albany State College. I said, "Evacuate immediately." I was right. The flood damaged twenty buildings at the college, many of which were almost completely submerged.

MARBURY ENGINEERING BEGINS AGAIN

Spec Dozier, one of my outstanding clients, also called to ask if his neighborhood in Merry Acres Subdivision would flood. Based on projected flood elevations, I recommended they evacuate. I was wrong. Floodwaters never did any substantial damage to that neighborhood.

Various governmental agencies calculated estimated high-water elevations and provided our company with the information. We knew a few days ahead of time the estimated high-water elevation for Baker County. Stephen Hart lived in that county, and I told him to take a survey crew and locate exactly where to expect flooding. One mobile-home park was in serious danger. Stephen warned the residents and had them move their mobile homes to higher ground. Most heeded his warning, but one did not.

Stephen pleaded for the owners of that mobile home to move. They refused. He warned them again. They refused. Then, in frustration, Stephen took a marking pin and drew a line near the top of one of the windows.

"The water will rise to that location," he told them. They still refused to move.

The floods came. Those who'd heeded Stephen's warning saved their homes. Water rose to within one inch of the mark Stephen had placed on the one mobile home where the owners refused to move. They lost their home to the flood.

At that time, I served as the Columbus, Georgia, Stake President for our Church, the Church of Jesus Christ of Latter-day Saints. As such, I also served as the Regional Welfare Agent for our Church in that area. This allowed me to coordinate more than seven thousand volunteers from our Church as they arrived to help with this disaster. More than five hundred arrived during the first weekend. More than fifty-five hundred arrived the second weekend, and approximately seven hundred fifty helped the third weekend. A few others came to help on future weekends.

We unloaded most of the supplies at my office, located at 2334 Lake Park Drive in Albany, Georgia. Our church provided nine truckloads of food and supplies. These were the large eighteen-wheel truckloads. Supplies included fresh water, food, work equipment, and medical supplies. They also included shovels, wheelbarrows, axes, saws, generators, water coolers, extension cords, bleach, ladders, and other items needed for immediate use.

We opened a Red Cross Shelter in our church building for displaced families. I stayed there much of the time to help coordinate relief efforts and ensure that the families staying there had needed food and water. We also provided medical personnel in case of medical emergencies.

My daughter and her family lived on Hoover Street in Albany, Georgia. Their home was not in a flood area, and they offered to live with us and let a displaced couple live in their home until the floodwaters resided. The couple had been there only one day when they called us.

"We have never been in a flood before. How fast does the water rise? I see it coming down the street."

I assured them that they had nothing to worry about, since the home was well above the 100-year flood zone. A few minutes later, they called again.

MARBURY ENGINEERING BEGINS AGAIN

"We're getting out of here. Water is rising really fast."

My wife and a few others immediately drove to our daughter's home. When we arrived, we understood what they meant. Water was already a foot above the floor elevation of the home. Quickly, several of us began moving what low-lying articles we could out of the home or to higher elevations. We moved her piano to the top of a couch at an elevation I told them was certainly high enough to protect the piano. I was wrong again. Water rose about three feet above the floor level of her home, flooding the lower part of the piano.

Since we did much work in polluted areas, we recommended all workers be vaccinated against possible tetanus. The local health department had plenty of vaccine but only two-inch needles. They asked me if my tetanus shots were current, and I replied that they were not. They grinned.

"We need some shorter needles," they told me. "We insist you allow us to give you your tetanus shot. If your church can supply shorter needles, we will use those. Otherwise, we will see how you like having your tetanus shot with the longer needles."

I wasn't sure if the longer needles hurt any more than the shorter ones, but I never liked shots of any kind. I wanted them to use the shortest needle possible on me. It took a lot of additional paperwork and approvals to allow needles of any kind to be shipped in large quantity. I was persistent, however, and the needles arrived quickly. The nurse took special pleasure in administering my tetanus shot. She told me the health department was sure I would find a way to get the shorter needles. They were right.

Well contamination was a problem to many homeowners. Our church sent dozens of bottles of Clorox for us to use in well decontamination. I organized several well-decontamination teams, instructed them on proper decontamination methods, gave them the needed Clorox, and sent them on assignments. Our teams decontaminated most of the wells in the area in just a few days.

The floodwaters created more problems than just flooding homes and businesses. They caused limesinks in addition to the well contamination mentioned earlier. They caused at least one other unexpected problem.

One of the homes on a small lot near the river was extremely well built, with a solid wooden floor. In fact, the entire structure was tightly put together. The floodwaters lifted the entire home, moved it from its existing lot, and placed it neatly on the adjacent vacant lot.

When the floodwaters subsided, the homeowner returned to view any damage to his home. It was gone. His lot was vacant. He walked over his lot, viewing with dismay the grass, trees, and small puddles of water still remaining on his lot. His mailbox still remained, although wet, but there was no house.

He looked to his right. There, in the middle of what was once a vacant lot, stood his home. Real estate law says that the person who owns a lot owns all structures located on that lot. Since this family's home was now on a different lot from its original location, the question existed, "Who owns the home, and who owns the furniture and other items inside?" I never found the answer to that question.

MARBURY ENGINEERING BEGINS AGAIN

I am grateful for the outstanding staff I had at Marbury Engineering during those days. All of us worked at no cost to the families and individuals we served during that period. Many worked longer than their usual hours and on weekends with no pay, other than their usual weekly salary.

HOSE BIB AND WATER FAUCET

As stated earlier, most reviewing agencies have knowledgeable and reasonable individuals—but not all. One winter I submitted a set of construction plans for a small commercial development. The design was a simple one, and I expected approval in a matter of one or two weeks. I was shocked to see my plans returned "Not Approved." There was only one comment regarding the reason for disapproval. The regulations required three water faucets in various locations on the site, and the reviewing agency said I had none.

I called the reviewer for an appointment. I explained that, in fact, I had shown four water faucets on the plans. They were labeled hose bibs, which is the same as water faucets. The reviewer was not satisfied.

"The regulations require water faucets, and you show no water faucets," said the reviewer.

"I showed four hose bibs. That is the same thing as water faucets."

"You did not show any water faucets, and the regulations say you must show water faucets."

"But hose bibs and water faucets are the same thing," I said again.

181

"If you expect to get these plans approved, you will show at least three water faucets."

I took the plans back to my office, changed the words "hose bib" to "water faucet" and resubmitted the plans. They were approved immediately.

DOUG WINGATE RECOGNIZES SERVICE

I always felt it was my job to treat clients' needs the same as my own. That was the way I felt about the need to improve my designs at Stonebridge Golf and Country Club. Tom Fischer, with the U.S. Army Corps of Engineers, contacted me about some wetlands at the club. He said we needed to do some work quickly, otherwise, the wetland permits would expire. Doug Wingate, the owner, was not available, but the work needed to be done immediately.

I proceeded to do the work without specific approval. I also did some other work that needed to be done in the development. My cost was more than expected, and I was reluctant to invoice based on our normal billing rate. Instead, I billed Doug an amount less than my actual cost. I felt the work was worth at least the amount billed and sent him the bill. A few days after Doug received the invoice, he came to my office.

"I will not pay this bill," he stated emphatically. "I didn't authorize the work, and I will not pay for it."

I acknowledged that he was correct and told him I would void the bill. I also gave him a copy of our expense sheet on the project. I asked him to review the information so he would know I had not tried to take advantage of him. I also explained that I felt I was looking after his best interests. If he felt differently, he owed nothing, and I would continue to work for him as long as he wished. He took the papers and stormed out of the office.

A few days later, Doug called, asking to see me. We set up an appointment, and Doug arrived as scheduled. Doug brought the papers I had given him and handed me a check. The check included not only the amount invoiced but also an additional amount to cover my cost of doing the project.

I told Doug he'd paid too much. He responded that he'd reviewed our cost and did not want me to lose any money over efforts to help him. He said it was unusual for people to look after his interests above their own, and he wanted me to know how much he appreciated it.

ROD HUTCHINSON AND THE ROWDY CLIENT

Rod Hutchinson was a large man and an outstanding designer. The great thing about working with Rod was his common-sense design and his ability to work quickly and efficiently. He always wanted things done correctly and had little patience for those who knew what should be done but refused to do so. He also expected others to act respectfully.

One day, we were working on a project that was managed by a project management firm I will not name. The project manager was an easily excitable, small individual prone to using improper language. On this particular day, he entered the office demanding to see Rod. He must have been in a foul mood, because he began the conversation in rather loud and demanding tones.

Rod was patient at first but soon became frustrated. The project manager continued ranting and raving about things he said should be done that weren't being done—at least not the way he said they should be done. The project manager's voice grew louder and louder until it could be heard throughout the office.

Rod decided the discussion had become too rowdy and asked the project manager to step outside so they could discuss the matter away from the rest of the office. Rod spoke in a quiet voice, but anyone could tell he was not happy. The project manager looked up at Rod. Rod was a towering figure of a man. He then considered his own size. He was a small wimp by comparison. The project manager said not a word but rushed out of the office, jumped into his car, and drove away quickly.

About an hour later, I received a phone call.

"Your man, Rod, asked me to go outside so he could beat me up. He needs to be fired," said the project manager.

"He was only trying to get you outside so that you would not disturb the rest of the office," I replied.

I never heard much from that project manager again. I believe he was fired, but I'm not sure. I do know I appreciated Rod taking care of a troublemaker and often wished I had given him a bonus for what he did.

QUALITY SEMINARS

In 1995, I assisted in the writing and preparation of the Quality Assessment Workbook published by the American Consulting Engineering Council (ACEC), later to be known as the American

MARBURY ENGINEERING BEGINS AGAIN

Council of Engineering Companies. Previous to that publication, I'd served as the national chairman of the Quality Management Committee for the same organization. Later, I gave seminars across the country on quality management.

My seminars reflected lessons learned through years of successes and failures. Some of the ideas taught were as follows:

- The interests of the organization are inseparable from the interests of staff.
- Understand fully before making decisions.
- Determine your leadership style based on individual needs and performance.
- Sustained improvement is best obtained through positive reinforcement.
- The best results come from integrity, courtesy, and kindness.
- Obtain commitment before implementing change.
- Keep the main thing the main thing.
- Teach and model correct principles. Correct behavior will follow.
- Do not let bureaucracy take the place of good judgment.
- Do not mistake activity for achievement.
- Problems are best solved before they occur.
- Avoid getting offended by trivial events.
- Leaders fail most often due to pride, arrogance, and out-of-control ego.
- Create an environment that makes it easy to do the right thing.
- Understand before you act.

- Fix the problem, not the blame.
- Listen – Listen – Listen.

I must admit I often failed in following many of the principles I taught, but not deliberately. I believed in, and still believe in, each of those principles. I still work hard at improving by following those principles.

LOWE'S FIRE OF 1996

On April 16, 1996, employees at Lowe's retail store in Albany, Georgia, noticed a small fire in a pallet of packages. Employees were unable to extinguish the fire, and it quickly expanded. The fire overwhelmed the automatic sprinkler system and destroyed the entire 85,000-square-foot building in minutes. The fire department arrived quickly but was unable to save the building or any of its contents.

A few days later, I received a call from Lowe's management, requesting help. They employed our firm to help locate another site, prepare a boundary and topographical survey of the new site, assist in obtaining rezoning, prepare the site design for the new building location, and obtain approval from all governmental agencies. That included the City Engineering Department as well as the Planning Department and City Commission. Not only that, they wanted all work done in approximately one month so that they could start rebuilding immediately.

I told them it could be done if the city officials agreed to expedite all plan review. We made no request for any leniency, just expeditious review. I also told my staff that we would have to work in shifts so that we could spend twenty-four hours a day working on this project. They agreed. I also had to contact other clients with deadlines and request that we have an extension to complete their work in order to meet Lowe's needs. My clients all agreed.

The fact that we completed the project on time is a tribute to not only our staff but also to all the public officials. As best I can remember, we worked two twelve-hour shifts. One shift came to work at seven in the morning and worked until seven in the evening. The other shift worked from seven at night until seven in the morning.

City officials were no less diligent. I submitted plans on one afternoon, and, immediately, the planning commission called a special meeting to review and approve our plans. The City Engineering staff worked late into the night to review our design and have comments back to us the next day. The City Commission reviewed and approved rezoning requests as quickly as regulations would allow. We worked each day with the hope that the City Commission would approve all rezoning applications, since we had no time to wait for approval if we were to meet all deadlines.

Lowe's also did their job. No employee went without pay, although some had to go to other cities and work until completion of the Albany, Georgia, store. Lowe's also responded to every request by us within twenty-four hours.

They paid us promptly. I'd explained at the start that we would have to charge much more than our regular rates due to having to work overtime and through the night. They told me that cost was no object, since they were losing more than one million dollars in revenue for every weekend of construction delay. We charged extra only for the additional cost to our firm in paying staff for the overtime work. We felt this was a community effort, and we were grateful to be able to do our small part in getting this needed facility back in operation.

All design met deadlines. All governmental approval met deadlines, and all construction met deadlines. It is a good feeling to be able to work in a community that comes together in times of need.

JOINT VENTURES

It was always exciting to work with other firms. My career included forming and working with Albany Design Associates, Marbury Ritter Scott and Turner, Marbury and Ritter, Marbury Associates, Logan Armentrout & Marbury, and Armentrout Marbury and Associates. There were also others.

My dad had helped start Albany Design Associates, which included our civil engineering firm, an architectural firm, a mechanical engineering firm, and a structural engineering firm. This was a separate corporation established to compete with larger companies for state and federal projects. With the death of Dick Richards, the architect head of Albany Design Associates, that corporation ended. Over the years, we formed other corporations in which our separate firms worked together.

It was interesting how we would sometimes compete against each other. Once we had a joint venture consisting of my dad as our firm's representative and Lou Lindsey as the structural engineering firm's representative. Another joint venture had me as our firm's representative and Guy Ritter as

MARBURY ENGINEERING BEGINS AGAIN

the structural engineering firm's representative. Guy Ritter and Lou Lindsey were with the same structural engineering firm. Of course, dad and I were with the same civil engineering firm. We often competed against each other.

Once we received a request for proposal from Dougherty County, Georgia, for the design of a Dougherty County Jail. More than twenty teams responded. Our firm was part of a joint venture with about fifteen of the twenty teams. When the list narrowed to five firms, the final five were invited to interview for the project. Our firm was part of the design team for all five firms, and I was our firm's representative. I also attended all five interviews. During the last interview, the head of the selection committee commented on my being at all five interviews.

"You're pretty sure about getting this job, aren't you?" the committee head asked.

"I'm really trying," I responded.

All firms knew we were part of several teams, and we always made sure everyone knew what teams we were on, to avoid any conflict of interest. Still, many selection committees over the years teased us about competing with ourselves when marketing for engineering projects.

With the help of Gary Harrell, Marbury Associates Inc. began on January 17, 1992, and continued until it was dissolved on November 9, 2002. The corporation consisted of Marbury Engineering Company as the engineering part of the corporation and surveying firms in various communities becoming associates. The first surveying associates were Bryson Langford in Shellman, Georgia, and J. B. Faircloth in

Cordele, Georgia. Later, Stan Folsom from Valdosta, Georgia, and other surveyors became part of our corporation.

When we contracted for any project, each company would agree on a specific scope of work and a specific payment. This prevented any controversy regarding how much each company would get paid for projects. During our almost eleven years of working together, we never had any arguments or disagreements regarding finances.

Stan Folsom was part of our Marbury Associates joint venture for several years. I remember one project in South Georgia where Stan and I surveyed part of the same property, each arriving at the final point of survey from different directions. Since both surveys consisted of traversing several thousand feet of survey, I expected our survey to come within about one or two inches of Stan's final point. When our survey met with Stan's survey, we missed his point by about one hundredth of a foot. That is less than one eighth of an inch. That was confirmation that both of our surveys were correct; however, I thought I would have some fun with Stan.

I called Stan's office, and he was not there. I left word that I'd found a terrible error in his survey and that he needed to contact me immediately. He got the message late, and it worried him so much that he called me that evening.

"Ritchey," he said. "What's wrong with our survey?"

"You made a terrible error," I replied.

"What was it?" he said.

"We did another survey where we checked the accuracy of your survey and found you missed the correct point by about one hundredth of a foot."

MARBURY ENGINEERING BEGINS AGAIN

"*How* much?" said Stan.

"One hundredth of a foot," I replied.

Stan laughed.

"You really had me going. I thought I had a big mistake," said Stan. Then Stan told me that, actually, I just had an error in my survey and hoped I would be more careful next time. Stan and I had many good times working together and still enjoy our relationship today, even though Marbury Associates dissolved many years ago.

PUDDLE EXPERT

Fonda gained the title "PE" from traveling with me on inspection trips as we inspected paving in parking lots. No, she is not, and never was, a professional engineer. In Fonda's case, "PE" stands for Puddle Expert.

I worked on many parking lots over my years as a civil engineer and professional surveyor. After contractors pave parking lots, the lots need to be inspected. One of the best times to inspect is after a heavy rain. Fonda and I would go to the sites and look for puddles. If the lot had been paved correctly, there would be no standing water. If there was standing water, that meant there were deficiencies in the paving that needed to be corrected.

Fonda became efficient at finding standing water and would advise me of the problems. I recorded the problem and notified the contractor. After the contractor corrected the problem, we would either wait for another rainfall or open fire hydrants and let the water flood the parking lot. If we found no standing water, we approved the project.

191

As Fonda went on more projects with me to look for puddles, I began to rely on her to locate puddles. In fact, I would call her my Puddle Expert, a real PE.

INLETS WITH NO OUTFALL

John Sperry, another outstanding civil engineer, once told me about a complaint by one of his clients regarding drainage near his property. His client felt his property should drain even though the property was low and that, therefore, water ponded on his property with every rainfall. John explained that water would always seek the lowest level, but that didn't satisfy his client. His client wanted no standing water on his property even after the heaviest rainfall.

After considerable discussion, John's client said he had a simple solution. The solution was just to put one of those inlets on his property and let the water drain into it. It didn't seem to matter that no outfall storm sewer was near the property. John's client felt that if an inlet were there, the property would drain—outfall pipe or no outfall pipe.

John often laughed about such a ridiculous statement. I, also, enjoyed hearing about this story. Later, it happened to me.

My son, Rick, and I had an engineering office in Boise, Idaho, from August 1997 until January 2002. A client wanted us to design a parking lot in downtown Bend, Oregon. Since I was not familiar with the storm drainage regulations in Bend, I called the city engineering department. I spoke with a man who worked in that department.

"What amount of rainfall do you design for in Bend, Oregon?" I asked.

MARBURY ENGINEERING BEGINS AGAIN

"What do you mean?" he replied.

"I need to know how to size the outfall storm drainage pipe in the parking lot. What size rainfall storm event should I design for?"

"Oh, just put a couple of inlets in the parking lot," he replied.

"Where are the outfall storm sewers located?" I asked.

"We don't have any. Just put a couple of inlets in the parking lot."

By now, I was really frustrated. "How much rainfall should I design for?"

"One or two inches," he replied again. "Just put a couple of inlets in the parking lot."

"Do you want me to design for one or two inches?" I asked again.

"Alright, design for one inch," he replied.

"Do you mean one inch an hour or one inch a day?"

"I mean one inch a year. We don't have much rain here. The soil is very sandy, and so we really don't need any outfall storm sewers. Just put a couple of inlets in the parking lot."

Bend, Oregon, actually has an annual rainfall of about eleven inches, but that is still an average of less than one inch a month. That is typical of the high desert, which is where Bend, Oregon, is located.

I designed the parking lot with two shallow inlets that contained no bottom. The rainfall drained into the inlets and then soaked into the sand at the bottom of the inlet.

Had John's client lived in Bend, Oregon, his solution would have worked perfectly. Of course, if John's client lived in Bend, Oregon, his property would never have flooded, since the little rainfall that might have fallen on the property would have immediately soaked into the sandy ground.

After four or five years, I decided it was too difficult to maintain an office so far from home and closed the Boise, Idaho, office. Rick moved to Marietta, Georgia, and we opened an office there for a short time.

Chapter 7

LIFE AFTER MARBURY ENGINEERING COMPANY

EMC ENGINEERING
I ATTENDED A SEMINAR IN Florida in 2003. That was where I first met Chuck Perry. Although I had no plans to retire, I did feel a need to provide for my staff when I was no longer able to manage Marbury Engineering. At age sixty-five, it was time to plan for the future.

Chuck was a principal with EMC Engineering and approached me about selling my company. We talked about the benefits. I looked at other possibilities and decided to sell to EMC Engineering. I did have other offers for more than what EMC Engineering offered, but I liked the EMC Engineering structure. We hired a firm to appraise both companies and agreed on a price.

On March 29, 2004, I sold the assets of Marbury Engineering Co. to EMC Engineering. The terms of the sale required me to continue with the firm for not less than two years. I stayed five years before retiring from that company. We had some good times together.

Shortly after EMC's acquisition of our assets, Cary Reed and I visited the home office of EMC Engineering in Savannah, Georgia. We walked into their offices and into the office of Charles W. Tuten, their director of surveying. Charles was a slim man, with a smooth complexion, and dressed in a suit with a white shirt and tie.

"You ain't no surveyor," said Cary.

I had the same impression. No surveyor I had ever worked with went around wearing a shirt and tie on regular business days. How could one go into the field looking like a New York City banker? Later, I learned that EMC Engineering had a dress code and that their managers dressed more formally in Savannah than those of us in southwest Georgia.

On August 29, 2005, Hurricane Katrina made landfall in Louisiana as a category-three storm, with winds of 125 miles per hour. Later it intensified to a much stronger hurricane, and the EMC Board of Directors agreed to let a few of us go to Louisiana to help. They donated a generator for us to take to some needy family.

We drove to Slidell, Louisiana. Heavy rains there totaled approximately fifteen inches. Most of the city was in ruins, but the rains had subsided by the time we arrived. Power was gone in most areas, and the hospitals were full. Many needing help were able to do only what they could in their own homes.

One lady had a severe medical condition that required her staying in a cool, comfortable environment, but her home had no air conditioning due to the lack of electricity. We told

LIFE AFTER MARBURY ENGINEERING COMPANY

HURRICANE KATRINA

her family we had a generator to donate and that she could use it to run her air conditioner as well as other appliances. Steve Hunter, an engineer with EMC Engineering, along with others, helped set up the generator, and soon her window air conditioner cooled the room. Later, the family told me that that generator probably saved her life.

On September 30, 2005, I scheduled another trip to Louisiana, this time to New Orleans. I never made it. While loading the car, I heard a scream from inside my home.

"Come quick—Big Momma is choking."

"Big Momma" was what we called Fonda's mother. She had Alzheimer's disease, was eighty-eight years old, and had been living with us for the past eleven and one-half years. I abandoned my plans to go to New Orleans and quickly called nine-one-one. We rushed her to the hospital, but she died on the way. She died of a heart attack.

My five years with EMC Engineering taught me many things. EMC Engineering, at the time we started working together, had almost two hundred employees. Marbury Engineering had fewer than twenty. The new culture required getting used to. I had always been able to make immediate decisions and proceed. Significant decisions at EMC Engineering required approval of their board of directors, a process often taking a month or more.

There was much more emphasis on making every job profitable. EMC monitored every project and every office for production and profitability. They based salaries and bonuses on production, jobs marketed, and profitability. I was not used to such accountability.

LIFE AFTER MARBURY ENGINEERING COMPANY

There was also more travel. At Marbury Engineering, I could get to work in about three minutes. I lived only about a mile from the office. EMC's headquarters was in Savannah, about four hours and about two hundred miles from my home. I drove there twice monthly for the first year or two. Sometimes I would go home from work asking myself, "What have I done?"

Overall, the decision to sell Marbury Engineering Company's assets to EMC Engineering was a good one. The benefits far exceeded the disadvantages. I made many new and good friends, friendships that continue to this day. Being one of the largest owners, I still had some responsibility for meeting payrolls; however, it was not the same as when I had the business alone. EMC had a staff of bookkeepers and many owners. We all shared in the cost of operation, and we all shared in the responsibility for collecting invoices.

Another advantage was the large pool of talent. If there was some aspect of a job I didn't know how to do, EMC Engineering had many additional staff members available for assistance.

When projects got behind schedule, backup staff was there to help. When a key principal was sick or out of town, other talented engineers came to the rescue. We also were able to provide a wider range of services, allowing us to qualify for a wider range of projects.

My income increased. With the larger staff, we had a marketing department that could go after larger and more profitable projects. The larger organization made it possible to provide more benefits. Although Marbury Engineering Company

199

provided excellent medical benefits, EMC Engineering provided not only the medical benefits but also more vacation, more holidays, and more sick leave.

EMC also provided more training. They often had "in house" seminars, including financial training. They held annual meetings where they discussed ideas like mission statements, governing values, and ways to improve productivity.

I retired from EMC Engineering after working there five years. During my five years, I helped start offices in Columbus, Georgia, and Valdosta, Georgia. Both offices continue to do well. I retired, not because I wanted to stop working, but because I felt it was time to give the younger generation a chance for leadership.

Working with EMC Engineering was a great experience. They always treated me well, and we continue to be friends to this day.

SRJ ENGINEERING

Mackey Saunders approached me after I retired from EMC Engineering about starting an engineering office as part of his architecture practice. He was the senior partner with SRJ Architects and planned to start SRJ Engineering. I would be the president and run the office until Mark Stalvey, a young engineer, could get his license. I would then retire from SRJ Engineering, and Mark would take over as president. Within a year, Mark took his exam and received his license. I retired from SRJ Engineering shortly after that.

Working with an architecture firm was a new experience. Mackey and I were longtime friends and had worked together many years. My firm usually did his civil engineering and surveying work, so we were familiar with working together. The main difference this time was that I had no staff to do much of the AutoCAD work. I needed to do much of the AutoCAD drafting myself. I was not very efficient.

I never fully appreciated the value of CAD professionals before. I learned one version of AutoCAD. Then AutoCAD upgraded the program. The upgrade was meant to simplify work and provide ways to increase production. To me, what it did was require me to learn a new system even before I fully understood the old system. "Simplify" meant "complicate," and "increased production" meant more study to learn how to do things I already knew how to do before the "new and improved" way of doing things.

SRJ did, and still does, outstanding work. One of their strong points is their emphasis on quality. They check every design several times for accuracy. They read every word of every specification to be sure it applies to the specific project. They do their own construction inspections and even use specialists to do construction inspections in addition to inspections done by the project managers.

The one thing I missed while working at SRJ Engineering was the ability to do survey work. We contracted with various surveying companies, including my old company, EMC Engineering, but did not have survey crews ourselves, so we were not able to do our own surveying. We did rent some GPS equipment for an out-of-town survey, and I enjoyed that experience. Other than that, however, we contracted our surveying work to other firms.

201

Shortly after Mark Stalvey received his professional engineer's license, I resigned from SRJ Engineering. Mark took over as president and did a fine job. My time at SRJ Engineering was short but pleasant and enjoyable.

RITCHEY MARBURY PE RLS

For one year, January through December 2011, I worked for myself out of my house. I had a twenty-foot by twenty-foot room to use as an office. My laptop computer sat on an old wooden desk, once used by my grandfather. I purchased a cheap plotter capable of plotting twenty-four-inch by thirty-six-inch drawings, purchased a supply of twenty-four-inch by thirty-six-inch plotter paper, and I was in business.

Clients were kind, and I immediately had a backlog of work, both engineering and surveying. Since I had no surveying equipment, I often subcontracted with Clay Miller, a local surveyor, who also worked for himself out of his house. Clay had the necessary surveying equipment and helped out with surveying projects.

One of the first projects Clay and I worked on together was an ALTA survey in Warner Robins, Georgia. This is a type of survey that must meet requirements of the American Land Title Association. Clay and I made up the entire survey crew, and our billing rates were typical of rates charged by registered professional land surveyors. In other words, our billing rate was approximately double the standard billing rate of a typical survey crew. In order not to scare the client because of our unusually high billing rate, we quoted the job

LIFE AFTER MARBURY ENGINEERING COMPANY

as a lump-sum project. In other words, we contracted to do the project for a fixed fee.

Our expenses on this survey proved that hourly rates are not always the best indicator of what is most economical. We priced the survey based on prices we would have used had we done the project with a typical survey crew. Although our combined billing rate was approximately double the standard billing rates for a typical survey crew, we completed the project a day early and under budget. I believe the reason was that both of us knew exactly what to do and immediately did it, with very little conversation between the two of us. I really enjoyed my time working with Clay.

Working from home had several advantages. Overhead was almost nothing. I could work whatever hours I wanted, and I had more time to spend with my wife and family. To me, however, the disadvantages outweighed the advantages.

First, working from home made it difficult to stay focused. I got not only the usual interruptions from inconvenient-but-needed business callers but also interruptions from personal phone calls to my family or myself. I had to do all my own bookkeeping, billing, collecting, and other administrative tasks. When I had worked with Marbury Engineering Company, EMC Engineering, and SRJ Engineering, all this was done for me. Now I had to do all this myself, which made it difficult to focus on the profitable engineering portion of my work.

Most of all, I missed the companionship of coworkers. I got lonesome. True, my wife was at home, and I liked that. Working by myself, however, with no one to double-check

203

my work and no one to talk with about various engineering solutions, was difficult. I would draw on the computer and then look up at blank walls.

My wife, Fonda, was wonderful. She hung pictures on the walls in my home office. She hung my college diplomas on the wall, and she hung my engineering and surveying registrations on the wall. At least that gave my home office some atmosphere.

Still, I would sometimes think about various engineering solutions and look over my shoulder to ask another's opinion, only to find no one there. I sat alone in an office with four walls and no people. I was confident I knew how to do the work. I just wanted to talk to someone about various alternatives. The highlight of the day came when Fonda entered the room. We would talk about our children. We would talk about our grandchildren or how cluttered the office looked.

Working at home involved many interruptions. Often the house phone would ring, and I would answer.

"Fonda has a hair appointment at 2:30 this afternoon."

"Hello, you have been selected to receive a free vacation."

"This is a brief survey and will take only a few minutes of your time."

Working at home just wasn't for me. I wasn't productive.

CITY OF CORDELE, GEORGIA

Although my daughter, Mitzi, and Steve Hunter, had divorced a few years previous, Steve remained, and still is, a good friend. Steve knew of the need for a city engineer in Cordele, Georgia, and mentioned the opportunity to me. At first, I thought the drive of forty-five miles to Cordele and another

LIFE AFTER MARBURY ENGINEERING COMPANY

drive of forty-five miles back, was too much. However, I liked Cordele and talked to Fonda about the job.

Around the middle of December 2011, I decided to call Jeff Johnson, then the city manager of Cordele. I interviewed with Jeff the next day and was hired the following day. This began a new experience. After working sixty-two years with private consulting surveying and engineering firms, having started at age eleven, I would now be working as the city engineer for a governmental organization. I wasn't sure if I would like it but decided to give it a try.

I had worked for myself or my family since 1949, sixty-two years, and suddenly I was no longer my own boss. I also was no longer working for a private consulting firm. For the first time in my life, at age seventy-three, I was starting a new career as a government employee. I was now the city engineer for the City of Cordele, Georgia.

My first office location was in the main building of City Hall, but soon I moved to the Public Works office. My supervisor was a man named Steve Fulford. As of the date of this writing, I have been working with Steve for more than seven years. Steve is an outstanding manager. What I like most about Steve is his judgment and common sense. Steve always listens to ideas from whoever presents them and generally makes decisions based on what is the most logical and in the best interests of the City of Cordele. He cares deeply for the people working with him and constantly strives to provide a better working environment for those on his staff.

Another attribute Steve Fulford has is his ability to insist on things being done correctly without micromanaging. Steve

205

tells me what he needs done, tells me when he needs it done, and lets me decide how to do the job.

One of the hardest things for me to do when I first started working for the City of Cordele was to do my own AutoCAD drafting. Although I did some limited AutoCAD drafting while working at SRJ Engineering and while working for myself out of my home office, for most of my professional life I had CAD professionals working for me. I would sketch what I felt was a good engineering solution and give the sketches to CAD professionals to produce finished drawings. This was no longer the case. I now had to produce the finished drawings myself, which meant I had to learn AutoCAD and become proficient at it. It took me approximately six months to gain a basic proficiency, but eventually I learned.

FIRST CORDELE, GEORGIA, ASSIGNMENTS

My first assignment in Cordele was to work with an individual named Nelson Barrett. Our primary job was to prepare the stormwater annual report. I enjoyed working with Nelson. He understood the stormwater ordinances and kept good records regarding what needed to be done. During the first year, we completed the annual stormwater report and submitted it to the state for their approval.

One morning, Nelson and I drove along one of the streets in the Pine Hill subdivision. Nelson stopped and called out to a young man walking along the street. He called the young man by name and told him how much he appreciated the young man's athletic ability. Nelson officiated basketball during times he was not working for the City of Cordele and recognized

LIFE AFTER MARBURY ENGINEERING COMPANY

the young man from previous games he had officiated. The young man smiled, and we continued on our way.

I appreciated the way Nelson took time to recognize a young man for no reason other than to make this youth feel good. That's one of the traits I find in many of the people with whom I work in Cordele. They all seem to care about each other.

Nelson left Cordele for another job in Atlanta, Georgia, and the stormwater technician position became open. A friend of mine, Brandon McGirt, was looking for a better job, and I told him about the open position. Brandon had worked for me many years previously and was an exceptionally bright individual. Not only was Brandon knowledgeable in the state of Georgia stormwater requirements, but he was an excellent CAD operator. Brandon was one of the most talented individuals I knew regarding computers. Brandon had previously worked with a computer company and understood how to program computers, repair computers, and actually do design work with computers.

I told Steve Fulford about Brandon's abilities and recommended that Brandon be considered for the job. Brandon applied, was offered the job, and accepted immediately.

Prior to Brandon's coming to work, I felt the City of Cordele needed an updated utility map, and I began working on a drawing of the utilities within the city limits of Cordele. Within about six months, I had a complete drawing showing the water lines within the city. Much of the work had previously been done by another engineering firm, and all I had to do was to upgrade their work.

207

As it turned out, the Cordele fire department needed a map showing the water lines throughout the city in order to meet some state requirements. The Fire Chief mentioned this in a staff meeting and said he realized it would not be possible to have such a drawing ready by the needed deadline. I smiled and told him I had already prepared such a drawing. He grinned and said he felt he could kiss me. I laughed and said, "A simple 'Thank you' would be fine."

Although I miss working with Nelson, I thoroughly enjoy working again with Brandon. Among other things, Brandon makes up for many of my shortcomings in the ability to draw using AutoCAD. If there is anything I can't do—and there are many—Brandon seems to know instinctively how to do it. Brandon downloads new software, removes viruses from computers, and seems to be able to solve almost any computer problem that occurs in the Public Works Department.

Shortly after Brandon came to work for Cordele, we purchased Global Positioning System (GPS) equipment. I had used this type of equipment previously and understood its value. This system allowed us to measure both horizontal and vertical distances within a fraction of an inch. Measurements were based on the equipment calculating our location relative to the location of satellites orbiting the earth. Surveying methods of measurements had gone from using chaining pins to using satellites.

One of the first jobs using this system was to survey an existing gas line in the southern part of the city. It ran several miles south into Crisp County.

LIFE AFTER MARBURY ENGINEERING COMPANY

Brandon and I worked together for the first part of the survey, but later, Jay Croxton worked with me to complete the survey. I was seventy-eight, and Jay was in his late twenties or early thirties.

So that no one would complain about my slowing them down because of my age, I worked as fast as I could, as hard as I could, and acted as though I never got tired. At the end of the day, we returned to the office. Jay dragged himself through the door, while I sprinted my way in, explaining what a fun day I'd had working in the great outdoors.

The office staff had a good time with Jay, telling me not to work him so hard. What they didn't know was how tired I *really* was and that I would completely collapse and fall asleep within two minutes of the time I walked into my home.

I must also confess that Jay was recovering from two hernia operations. He really should not have been at work at all, but Jay was a trouper, and he knew the importance of getting the job done as quickly as possible.

When I began working for the City of Cordele, I worked Monday, Tuesday, and every other Wednesday. Shortly thereafter, the water-and-sewer superintendent left, and Steve requested me to work full time in order to help Jimmy Jackson, the new water-and-sewer superintendent. I considered this a great opportunity, but I explained that I would not be able to work on Thursdays since I was the current president of the Albany Rotary Club. The Albany Rotary Club met on Thursday afternoons. Once completing my time as Rotary President, I proceeded to work five days each week, and still continue to do so.

It took me only a short time to realize that Jimmy Jackson was excellent in his role as water-and-sewer superintendent. Although I was able to help him with some computer applications and was able to do some engineering calculations in order to determine the proper size for water and sewer lines, I learned more from Jimmy than he ever learned from me.

GEORGIA MUSEUM OF SURVEYING AND MAPPING

On September 12, 2017, my wife's birthday, I noticed an article stating that the Georgia Museum of Surveying and Mapping, located in Warrenton, Georgia, would close in the near future unless it could find a new home. Dan Crumpton, a local surveyor in Warrenton, had started the museum and was now seeking someone else to house and manage it. The article stated that, unless they found a new home, it could be lost forever.

I immediately contacted the Thronateeska Heritage Center in my hometown of Albany, Georgia, to see if they were interested. They were. I had many historical maps, instruments, and other items that I could donate to the museum. I did.

In January 2015, the museum relocated to Albany, Georgia, and, after seven months of preparation, the museum reopened at the Thronateeska Heritage Center on September 26, 2015. It contained many items from my old office, Dan Crumpton's exhibits, and other mapping instruments, tools, and maps, some dating back to the eighteenth century.

LIFE AFTER MARBURY ENGINEERING COMPANY

CORDELE, GEORGIA, WATER METER SURVEY

After I'd worked a little more than five years for Cordele, Benny Harpe approached me with another opportunity. He wondered if I would have time to work with him to produce a map of all of the water meters within the city. Now that Brandon was doing most of the work in preparing the stormwater management plan, I felt like this would be a good opportunity for me to get to do some work outside of the office. There are almost six thousand water meters in the city, and our goal was to survey each meter and put it on a map. We would also prepare a spreadsheet showing the latitude and longitude of each water meter.

Surveying water meters was a fun experience. I loved working outdoors. Some days the temperature was in the nineties, and some days the temperature was in the thirties. I still enjoyed working outside, but I liked it much better when the weather cooperated.

We began the survey on May 2, 2017, and completed the field survey on May 1, 2018, exactly one year from starting. We field-located 5,961 water meters plus many fire hydrants and water valves. My final drawing showed only the 5,961 water meters and no fire hydrants or water valves, since it was difficult enough to show each individual water meter along with the street address. Showing water valves and fire hydrants would make the map so cluttered it would hardly be readable. I completed the drawings on September 14, 2018. The drawings totaled 111 sheets.

My goal was to complete the field survey before my eightieth birthday, which would be on May 18, 2018. We

211

finished a little more than two weeks early. Benny and I may be the only ones who can truly say that we have walked all the streets in the City of Cordele, Georgia.

Surveying all the water meters within the city provided me with the opportunity to observe living conditions for not only wealthy residents but also some of the poorest.

Some of the poorest people are the kindest. One hot summer day, Benny and I proceeded to locate water meters in an old apartment complex. In some places, the grass looked like it had not been mowed in months, and, in other places, the ground was so dry that not even five blades of grass covered the front yard.

During one of the first weeks of surveying, I failed to bring water with me. Tired and dizzy, I sat down on a small bench in front of one of the apartments. Shortly a lady approached me.

"Would you please bring me a glass of water?" I asked.

Two or three minutes later, she handed me a glass of cold water and wanted to know if I would like some more or a Sprite. She also offered the same to Benny.

A few days later, Benny and I surveyed the water meter at another house. The lady of the house seemed happy and relaxed. She appeared pleased with the flowerpot used to hold flowers during the springtime. Benny and I had to look twice to be sure the flowerpot was what we thought it was. Sure enough, the flowerpot was a toilet. Adorning a prominent place in her front yard was a white toilet, lid up, and plants growing out of the bowl. It seems everything has more than one use.

Dogs presented a different problem. When you walk every street in a city, it's amazing how many dogs you encounter.

Luckily for us, most of the aggressive dogs were either behind fences or on chains. No dogs bit us during our survey, but we did have one or two close calls. My scariest experience was when five dogs darted from an open door and ran straight at me.

Benny and I were surveying a small residential street in Omar Heights Subdivision when I heard dogs barking inside a small dwelling. The owner opened the door, and out jumped five dogs. They ran straight for me. I froze.

I would like to say I wasn't scared, but that would be untrue. I was not just scared—I was terrified. The first dog jumped up my left side and another jumped on my chest. I did not move. Then the other three dogs circled me on every side. I stood motionless. What else could I do?

After about a minute, I noticed the dogs were wagging their tails and nudging me from all directions. Cautiously, I reached down and petted the dog that acted the friendliest. Soon I was petting all the dogs. They were all friendly. What a relief. Shortly after, the owner appeared at the door and called the dogs back into the house. I don't remember Benny even commenting. He just considered dogs to be part of walking the streets of Cordele.

CORDELE, GEORGIA, HURRICANES, AND TORNADOS

On January 2, 2017, tornados ripped through my hometown of Albany, Georgia, and throughout Cordele, Georgia, where I continued to work as city engineer. Straight-line winds tore down power lines, trees, and roofs. One mobile home park in southeast Albany was almost totally destroyed. Several people died. Cordele had similar results, although not as severe. I was without power for a week, and the temperature was below freezing.

My wife's brother-in-law died the first of that year, and we drove to Cleveland, Tennessee, for the funeral. We returned home after the funeral to freezing weather. After spending one night at home, we found a hotel room at the Holiday Inn and stayed there three nights, until power returned to our home.

Again, on January 22, 2017, another tornado struck. The January 22 tornado did not affect our home, and we did not lose power during that event.

A year and a half later, Wednesday night, October 10, 2018, Hurricane Michael struck the same area. There were no reported deaths in either Albany or Cordele, and few injuries, but power outages were almost universal. Mitchel EMC, the power company serving the south part of Dougherty County, reported that almost one hundred percent of its customers were without power. Most traffic lights failed to function in both my hometown of Albany, Georgia, and where I worked in Cordele, Georgia. I was without power five days. Benny Harpe, a co-worker in Cordele, lost power for nine days.

We were fortunate. Hurricane Michael did greater damage and caused many deaths in other areas, such as Florida.

I learned from these experiences how so many people work together during natural disasters. When Hurricane Michael struck my home, we lost power, along with most of the community. We were fortunate to have a generator, but it quit working after a couple of days. The same day it quit working, Boyd McMurtrey and his wife, Sherry, brought another

LIFE AFTER MARBURY ENGINEERING COMPANY

generator to our home. Boyd and Sherry were friends from our church. I gave my broken generator to Benny Harpe in Cordele, and another worker in Cordele had it functioning in less than thirty minutes.

In the 2017 tornados and in Hurricane Michael, volunteers from all over the area came together to remove trees, clean debris, and provide food and shelter to those needing help. I personally worked with those volunteers on weekends and often took pictures to record the damages. I quickly realized I could not do in my eighties what I could do when I was younger, but I could still help.

Cordele lost power for weeks, and even streetlights failed to function. Traveling to work in the dark was a frightful experience. One day I saw three wrecks on my daily forty-five-mile drive to work.

I generally left for work at six thirty in the morning and arrived about seven thirty. With no streetlights, as I entered town, cars darted across streets at random. Sometimes cars would sit at all four street corners with no movement. Then cars would all move across the street together, only to stop again in order to avoid collisions. Somehow traffic continued to move, although slowly. Amazingly, every day I arrived safely to work and on time.

On Thursday, October 11, 2018, Cordele had no power in most of the city, and we were all told to stay home. The next day, I went to work at the usual time only to find no one there. Power was still off, and I returned home. Monday, when work finally resumed at the city, many locations were still without power, and also many streetlights failed to work.

215

HOME IN ALBANY, GEORGIA DAMAGED DURING THE JANUARY 22, 2017 TORNADO

LIFE AFTER MARBURY ENGINEERING COMPANY

I spent much of my day reviewing damage and assessing what was needed to restore the city to full operation. A few days later, I was back reviewing plans, inspecting construction sites, and making field surveys.

IN MY EIGHTIES, WORKING FULL TIME, AND LOVING IT

Sometimes people ask me why I don't retire. Why should I? I'm doing what I love. I'm working with people I enjoy being around, and I feel like I'm providing a service. What would I do if I retired? Probably just sit around, watch television, and twiddle my thumbs. Not much of a life. This way, I get to exercise my mind and body while having fun every day.

I get up around five forty-five each morning. Around six thirty I'm driving to work, arriving there about an hour later. During the drive to and from work, I listen to audiobooks. I now get to listen to the books I had always wanted to read in the past but never had time.

When I first started working in Cordele, I sometimes entered the building first—and set off the burglar alarm. After this happened several times, they gave me my own burglar-alarm code. Sometimes I worked late and found myself locked out of the building if I was working away from the office. I finally got adjusted to their schedule.

They also gave me a remote-control device that would electronically open and close the gate that separated our parking lot from the street. When I first tried the remote, Marcia Pridgen and Judy Woodard were leaving the parking area. They opened the gate with their remote and started to drive out.

217

Not paying attention, I hit my remote, causing the gate to close just as they approached. I watched the gate close. They backed up, looked at the gate, and pressed their remote again, opening the gate. I did not notice the gate open. I just saw it close and hit my remote again. Again, the gate closed on Marcia and Judy. When they finally got the gate open long enough to drive out, I just waited and then drove out myself. I did finally confess to what I had done—but only after I felt it safe to do so.

I love how the day starts in Cordele. When I first enter the building, around 7:30 in the morning, I usually greet Marcia Pridgen, who is already working. Often Steve Fulford, our Public Works Director, will arrive about the same time as I do and invite me to enter the building through the side door leading directly into his office. Other times, I enter the building through the back door. Usually there are several members of the Cordele staff already there, talking, joking, and waiting to start work.

When I sit down at my desk, the aroma of coffee fills the air with a wonderfully pleasant smell. I don't drink coffee, myself, but I love the smell of it. If the coffee isn't already prepared when I enter my office, Benny Harpe is usually there preparing it. If coffee is not made by eight that morning, Tina Bearden will have it made shortly afterwards.

Jay Croxton soon enters the room laughing and giggling as though he had just played some fun joke on another member of our staff. My office is in the conference room of the Public Works building, so I get to see most of the Public Works staff as they arrive for work.

LIFE AFTER MARBURY ENGINEERING COMPANY

Although Cordele provides me with a computer, I also always bring my own. I have software on my own computer that is not on the Cordele computers, which allows me to be more efficient with some projects. Being able to work with two computers at the same time allows me to get more work done in a more efficient manner. Brandon McGirt was always available to solve any computer problems that occurred, whether on my own computer or computers owned by the City of Cordele.

Some days, I spend much of my time working on AutoCAD drawings or preparing reports required by local, state, or federal agencies. Other days, I review plans submitted by consulting engineering companies. I usually complete plan review within the same week as submitted, and I have never failed to complete plan review within two weeks of receipt of plans.

Other days, I go into the field to do GPS surveys. Sometimes I do site inspections for construction work in progress or recently completed. One morning I watched Jim Faircloth as he worked with a drone pilot do aerial photography and a drone survey over the site of Harris Press in Cordele, Georgia.

When I first started working with counting pins and a survey chain, I never dreamed that someday I would do similar surveys with drones operated by computer or that I would be making measurements using satellites that circled the Earth thousands of miles overhead.

One thing I really like about working with the Cordele staff is their desire to assist rather than hinder. The people

219

I work with always seem to be more interested in fixing the problem than fixing the blame, and they are always willing to help.

Of course, I love working in Cordele. Of course, I love working in land surveying and civil engineering. Wouldn't anybody?

I plan to continue working for the rest of my life and hope to keep working in the next life, also. Retirement just isn't for me. I love what I do. I love working with the people around me, and, hopefully, I can make a little difference for good every day. This is a great life.

**DRONES USED IN SURVEY
OF
HARRIS PRESS PROPERTY
IN
CORDELE, GEORGIA**

ABOUT THE AUTHOR

RITCHEY MARBURY began his career in surveying and engineering in 1949, at the age of eleven, while working for his father on a surveying crew. Later he earned his Bachelor of Civil Engineering and Master of City Planning degrees from the Georgia Institute of Technology. After serving two years with the U.S. Army Corps of Engineers, he became registered as both a land surveyor in the state of Georgia and a professional engineer in the states of Georgia, Alabama, Florida, and Idaho.

In 1965, after his honorable discharge from the U.S. Army Corps of Engineers, he began working full-time with his father at Marbury Engineering Company. He served as Vice President and later as President and CEO of Marbury Engineering Company until the merger of his company with EMC Engineering Services, Inc., in 2004. For the next five years, he served as a Senior Vice President and West Georgia Division Director of that company.

After retiring from EMC Engineering Services, Mr. Marbury worked with SRJ Engineering and as a private consultant until he accepted a position as City Engineer with the City of Cordele, Georgia. At the time of this writing, he has been with the City of Cordele, Georgia, for almost eight years and continues working there. He loves his work and has enjoyed working with every company with which he has been associated.

As a past national chair of the Quality Management Committee for ACEC (the American Consulting Engineer's Council, now known as the American Council of Engineering Companies), he helped in the preparation of the Quality Assessment Workbook published by ACEC in 1995. He has conducted seminars on planning and leadership for various organizations. These include the China Association for Science and Technology in Beijing, China; the national meeting of the Consulting Engineers Small Firm Coalition in Denver, Colorado; the NSPE Professional Edge Conference in Houston, Texas; the American Consulting Engineers Council's annual meeting at Hilton Head, South Carolina; and the Hawaii Consulting Engineers Council in Honolulu, Hawaii.

Ritchey loves life and enjoys people. He hopes those reading this book can share many of his fun experiences.

www.ingramcontent.com/pod-product-compliance
Lightning Source LLC
Chambersburg PA
CBHW052022070526
44584CB00016B/1856